Kids Activities To Help Save The Planet

R.L. Gemeinhardt

Other Books By R.L. Gemeinhardt available from Amazon

- Earth Dwellers Guide To Recycling And Environmental Conservation For Kids And Teachers

- Earth Dwellers Guide To Recycling And Environmental Conservation For Companies

- Earth Dwellers Guide To Recycling And Environmental Conservation

- DEMOB THIS!!! A Jack Owens novel as he deals with the spill from hell!

- Texas Used Oil Management: A Practical Guide To Compliance

- Author site RLGemeinhardt.com

Text copyright © 2020 R.L. Gemeinhardt. Cover art copyright © 2020 R.L. Gemeinhardt. All rights reserved.

No part of this publication may be reproduced, distributed, or transmitted in any form or by any means, including photocopying, recording, or other electronic or mechanical methods, without the written permission of the copyright owner, except in the case of brief quotations embodied in reviews and certain other non-commercial uses permitted by copyright law.

Requests for authorization should be addressed to VRM Group, LLC
Attn: Ron Gemeinhardt 190 — B2 Gulf Freeway, #126 League City, TX 77573

Cover Design by Debra Gemeinhardt
Edited by Mich Fisher

ISBN: 9798665711096

Limit of Liability/Disclaimer of Warranty: While the publisher and author have used their best efforts in the publication of this work, neither author nor publisher makes any representations or warranties with respect to the accuracy or completeness of the contents of this work and specifically dis- claim any implied warranties of merchantability of fitness for a particular purpose. No warranty may be created or extended by sales representatives or written sales materials. The advice and strategies contained herein may not be suitable for your situation. You should consult with a professional where appropriate. Neither the publisher, nor the author shall be liable for any loss of profit or any other commercial or personal damages, including but not limited to special, incidental, consequential, or other damages.

DEDICATION

To my grandkids in a firm belief that any real change in the condition of our planet will only come about if we can get the younger generations involved and help them see the benefits of doing what we can as individuals and groups to help save this planet and even reverse the environmental destruction we are headed in as earth dwellers on this wondrous planet we share with all living things.

"Never doubt that a small group of thoughtful, committed citizens can change the world; indeed, it's the only thing that ever has."

— MARGARET MEAD

Table of Contents

	Page
Dedication	5
Preface	8
Introduction	10

1. WHAT IF WE DON'T DO IT! 14
   Direct Threats On Coral Reefs
   Indirect Threats On Coral Reefs
   Where Are They Located?
2. WHAT STOPS PEOPLE 20
3. WHY WE SHOULD 21
   What Does It Mean To "Recycle"
   Waste Related Things We Do
   Why Recycle
   Things To Do To Reduce Waste
   Education, Knowledge, Things We Need To Do
   What To Do Next
4. STARTING OFF WITH THE THINGS FOR TO DO AT HOME AND ALL THE FAMILY CAN PARTICIPATE IN... 36
   Plastic Or Paper Or Cloth
   Designate A Container In Your Room For Scrap Paper
   Plastic Rings On Drink Cans
   Aluminum Cans
   Use A Plastic Drinking Bottle = Not Disposable
     The Human Impact
     The Animal Impact
     The Environmental Impact
   Pack A Waste Free Lunch
   Donate Toys And Clothes
   Reduce / Reuse – Repair / Recycle
   Monitoring Your Water Usage
5. SOMETHING YOU SHOULD NEVER DO THAT REALLY HURTS THE ENVIRONMENT................... 71
   Trash Out The Window Or Just Anywhere
6. ACTIVITIES FOR THE 4-7 YEAR OLDS 75
   Homemade Puzzles
   Homemade Wind Chimes
   Toilet Paper Roll Car Craft

	Page
Egg Carton Turtle	
7. ACTIVITIES FOR 7-TEENS	91
Make Recycling Bins For The Home	
Recycled And Planting Containers	
Bugs, Lizards, Other Small Creatures And Plants In The Back Yard	
8. GROUP ACTIVITIES FOR SMALL OR LARGE GROUPS OF ALL AGES WITH ADULT SUPERVISION	102
Beach Litter Collection	
Plant A Tree	
Plant A Garden	
9. A REAL IMPORTANT ACTIVITY FOR KIDS AND EVERYONE "EARTH DAY"	117
Participation In Earth Day	
10. Conclusion	122
References	124
About The Author	134

PREFACE

This book is, "Kids Activities For Kids To Help Save The Planet" and have some fun at the same time. It is written for many age groups to read on their own and even do some of the activities by themselves but for age groups below about 10 they should be supervised by adults until you are sure they can do them safely. Kids are the most important people in our lives and as parents or guardians just like the teachers in their lives. We should do what we can to teach them about how important saving the planet is, and even as kids they can help. Without the participation from you and kids in your life our likelihood of any success regarding saving this planet is slim to none. The youth of our planet, as in any aspect of our survival as the human race, will rely on their participation in change and more than anything else in their understanding of the importance of what and why we need to keep doing or change what we do in our everyday lives to save this planet and keep it healthy. This book covers many activities focused on ones that are achievable by kids of various ages, some with the hands on guidance and assistance of adults but many that can be done by older kids all by themselves. It gives all involved recycling, reuse, or environmental conservation activities that not only can teach important lessons and ways to help save this planet but in many cases can save money and most importantly provides ways to get our children of the world involved. Every effort was made to insure the instructions are clear, and even step by step processes included where needed. With the activity listings a reference list is provided not for only the adults involved but the kids involved. Every effort was made to just plainly inform about many things in your life that can be done to make a difference.

As part of this human race living on this wonderful planet we owe it to each other and our future generations to insure the quality of our existence and the health of this planet is protected and we all owe a duty to the children in our lives to help that process of understanding and involvement at an early as possible age to motivate and even inspire our future champions of change to save and protect this planet.

My intent is a more realistic approach, and that is just to give the reader a starting point to JUST DO SOMETHING!!!

INTRODUCTION

"The Earth does not belong to us: we belong to the Earth."

— MARLEE MATLIN

I HAVE BEEN DEALING with environmental issues for more than 40 years which also means dealing with the negatives associated with some people not doing what they should be doing. Some people just don't take the time to understand or accept all the reasons why our planet is in the trouble it is in environmentally that impacts all our air, water, land, and all the plants and animals around us and in the oceans, lakes, ponds, forests, fields and deserts, anywhere where life exists in all its forms. This book is written for you to better understand the problems and give you some ways to help solve the problems in small and big ways. More importantly I hope it gives you some ideas to think about and talk about with your friends, your parents or guardian, and Teachers to discuss because the way we can change all the bad things happening to the planet and fix the problems are with you and the adults and teachers in your life really working hard on the problems.

I hope all the stuff in this book is in language as younger children you can read to gain understanding on many of the issues and the negative impacts that specific activities cause and could be prevented or at least mitigated, if done differently, but please don't hesitate to ask the adults in your life or the Teachers about any of the issues or do more investigating on your own. Most of all, I hope everyone who reads this recognizes that your impact on each other's lives is not just what you tell each other what should be done but show them by how you live your own lives as an example of what it means to be environmentally responsible in your effort to save our planet is really the most important issue of all.

As identified in my first book in this series my background is with the petroleum industry where I was a fuel specialist in the Air Force back in the Vietnam Era and then went on to work for Shell Oil company for almost 30 years. While in the Air Force I worked with several age groups doing volunteer work with the church and also worked as an instructor in a day care facility on one of the bases I was stationed at and worked in the parks and recreation department as a technical theatre director doing summer theatre with young people, worked as a student Teacher at a high school, and as a psychology undergraduate student worked in child development studies. The point of all this information is that I want you to know is how much I appreciate the youth in our lives and we must get involved to support them in their development and education of the important things in life like the arts and now, for so many years in environmental. How to work with each other to achieve common objectives in doing theatre or playing together and especially now in this battle all of us are in to save the planet.

I have provided a long list of references in support to the information provided with a lot of reliance on governmental and other public sources with a few academic references. I wanted to do this book, like the others I have done, where you could check out the information for more detail using your computer and other sources on many issues easier than using a lot of technical references or articles that would be difficult to review.

The difference in this book from the others is that KIDS can do all activities in some way either by themselves or with some help but of course the age range suggestions are just a suggested range and the children who can actually do the activity depends on the specific child or group:

A few environmental terms below are to help you understand what all this stuff is about. The first one is biosphere which has a scientific technical definition as follows:

Biosphere: The biosphere is the zone of air, land and water where organisms exist. It is commonly known as the global sum of all ecosystems and consists of several layers including the atmosphere, the lithosphere and the hydrosphere. But, a better understanding of the term, from my view anyway, especially for people who have not spent the time in all the books us environmentalists do, and as communicating is always part of the problem, and in my view, we try to put too much science in all this environmental stuff when plain English will do. The definition I like to use is from the Biology Online Dictionary (1) that I will use as we go forward is that the biosphere is the part of the earth where living things exist and that it is the part of the earth (or planet) that is capable of supporting life that includes all the plants, animals, birds, insects, fish and all the things that live in the water and of course humans everywhere.

Sustainability: is another term I would like to get you understand with the definition I would like you to be aware of as the Heart Scale also considers this factor as well: From the United States Environmental Protection Agency (USEPA) which is a government agency that is the most important one in the USA that is supposed to make sure our environment is protected and help everyone along with enforcing the laws defines the word as "Sustainability is based on a simple principle: Everything that we need for our survival and well-being depends, either directly or indirectly, on our natural environment (see more about the natural environment below). To pursue sustainability is to create and maintain the conditions under which humans and nature can exist in productive harmony to support present and future generations."

Natural Environment: All living and non-living things occurring NATURALLY, meaning in this case not artificial. ... This ENVIRONMENT surrounds and has interaction with all living things as climate, weather and NATURAL resources; soil/rock, air, water, sun all part of the physical environment that affect human survival and economic activity.

I will provide a Kid Age Range for each activity: It identifies the activity by a child age when required. No matter what your age you should make sure your parents or other adults in your life know about what you are doing if they are not already helping you with the activity. Adults, Teachers, please note that every minor child needs adult supervision but many activities can be done solely by children with minimal direction and in some of the activities, if of reading age, can be done on their own with some cautions and supervision as required given by adults when scissors or other cutting or hammering activities are required. Also some require glue or paint so please make sure non-toxic materials are always used without exception.

Note: Comments are included for clarification purposes on some of the kid ranges provided.

1

WHAT IF WE DON'T DO IT!

Okay so with any discussion about the good things you can do to help our planet you should know about why it is so important to all of us and all the living things on the planet. It is not meant to scare you but no matter what your age you should know that what we do in life matters but also know that things can change and if we all work together we can make everything better for you, the adults, and all the people, and the whole planet in the future.

Simply, the planet you and me live on is sick and getting sicker and if we don't start doing things that will help it get better it could die and make it real tuff to live our normal lives anywhere on this big planet, including all the other living things that will also suffer or die off if we don't do what we need to do as humans, who are supposed to know better. Adults are in a great deal of argument and different opinions based on their understanding of the science involved as to when this planet will die or at least when our lives will be impacted tremendously, but what is not an argument is that it WILL HAPPEN whether it is 50 or 500 years should not matter as to changing in what you and the adults do if you really care about this planet.

Impacts are many but a few of the numbers and impacts are as follows:

1. Every year, we take from the planet surface or below ground an estimated 55 billion tons of fossil energy, minerals, metals and bio mass (plant life) from the Earth.

2. The world has already lost 80% of its forests and we're continually losing them at a rate of 92,664 acres (375 km2) per day!
3. At the current rate of taking away trees (deforestation), 5-10% of tropical forest species will become extinct every decade.
4. Every hour, 1,692 acres (6.8 km2) of productive dry land become desert.
5. 27% of our coral reefs (places where a lot of sea-life live just off shore) have been destroyed. If the rate continues, remaining 60% will be gone in 30 years.

DIRECT THREATS ON CORAL REEFS

6. People visiting areas where coral is (Tourism): The simple act of standing on corals and touching them can cause damage that will take a thousand years (millennia) to mend, if ever. Boat propellers, snorkeling, dropping anchors on the reef and more are just some of the examples of how humans are harming the reefs.
7. People in business to catch sea life (Commercial) who use traps: Lobster traps are light enough to be moved by the current but heavy enough to damage a coral when it comes in contact with one.
8. Destructive Fishing Methods: Using chemicals that are bad that kill plant life in the coral area. Fishermen break the corals in order to get the fish that they stunned. People also bang the reef with sticks to get at the fish or even use dynamite to stun fish so they float to the surface but it also destroys the coral they live in. Even though people are not supposed to be doing this and it against the law they still do it.

INDIRECT THREATS ON CORAL REEFS

9. People living near the coral reefs that includes some farmers let bad chemicals, oil pollution, and stuff that normally goes in the toilet and is never supposed to get to the coral reef called sewage is allowed to get into the water. They are either dumped directly or carried by the rain or rivers into our oceans.
10. Too much soil that does not belong in the oceans called sediment: This comes from farms or construction sites near the ocean. The sediment kills corals by blocking the sunlight or clogging their mouths. (Yes, they have mouths)
11. Climate Change is another harm caused by us using fossil fuels (Gas, Diesel, Jet Fuel and lots of other stuff made from crude oil). Global warming is part of this climate change that has caused an increase in coral destruction, and it can only get worse. 80% of the added heat in our climate is being absorbed by the ocean. Corals cannot survive on waters with too high temperatures.
12. Pollution: Corals need clean water to live. The garbage that we throw in our oceans affects the quality of the water.
13. We have a garbage island (yes an island) floating in our ocean, almost unbelievable mostly comprised of plastics—the size of India, Europe and Mexico combined!

WHERE ARE THEY LOCATED?

- There are several garbage islands in the ocean and the most "popular" of them is the Great Pacific Garbage Patch (in the Pacific Ocean). The size cannot be pinned down since it's constantly growing. The island is made up of pelagic plastic, debris and chemical sludge and is a collection of waste coming from the surrounding countries, which are brought to this location by the currents.
- The Eastern Garbage patch can be found between Hawaii and California and scientists have estimated its size to be twice that of Texas!
- The Western Garbage Patch can be found east of Japan and west of Hawaii. These 2 patches are connected and they are the biggest plastic landfills in the ocean today.
- The Atlantic Ocean also has a garbage patch in the Sargasso Sea. Aside from the Pacific Ocean and the Atlantic Ocean, plastic landfills are also turning up in other major tropical oceanic gyres in the world.

14. We are using up 50% more natural resources than the Earth can provide. At our current population, we need 1.5 Earths, which we do not have.
15. 5 basic causes of the environmental problems we face today. Population growth, wasteful and unsustainable resource use, poverty, failure to include harmful environmental costs of goods and services in their market prices, and insufficient knowledge of how nature works.
16. Environmental Concerns. Our Mother Earth is currently facing lot of environmental concerns. The environmental problems like global warming, acid rain, air pollution, urban sprawl, waste disposal, ozone layer depletion, water pollution, climate change and many more affect every human, animal and nation on this

planet.
17. We as humans impact the environment in several ways. Common things we do include putting bad things into the water, increased pollution and greenhouse gas emissions by using fossil fuels, depletion of natural resources (coal, trees, other minerals) all add to global climate change.
18. Littering and Landfills: Littering simply means disposal of a piece of garbage or debris improperly or at a wrong location usually on the ground instead of disposing them into a trash container or recycling bin. Littering can cause huge environmental1 and economic impact in the form of spending millions of dollars to clean the garbage up—anywhere it is put.
19. Landfills on the other hand are nothing but huge garbage dumps that make the city look ugly and produce toxic gases that could prove fatal for humans and animals. Landfills are generated due to large amount of waste that is generated by households, industries and healthcare centers every day. While they can be used for other purposes if managed right, they still can represent problems and threats to all parts of our environment.
20. Billions of plastic bags are made each year. Of these bags, one hundred billion are thrown away according to Worldwatch Institute, with less than 1 percent finding their way into a recycle bin. The end result of this is around 1 billion birds and mammals dying each year by the ingestion of plastic.
21. The United States makes up less than 5 percent of the population on earth, yet we easily consume over 30 percent of its resources. While we humans would appear to be doing well, spreading our population like wild fire across the globe, the diminishing resources

and other life forms on the planet tell a different story. "We are in the midst of a mass extinction, an event not seen since the disappearance of the dinosaurs, 65 million years ago," says the Worldwatch Institute.
22. Americans dump 16 tons of sewage into their waters, every minute.

Okay, that is enough of the scary reasons we need to do something. A reference list is provided at the end of this book for more information and I encourage you to do some more investigation on your own. Just trying to make the point that the problems are real and unless you just believe that the scientists are all lying. Some of us scientists do disagree on some of the facts but that is okay because discussion and more research is so important to understanding any big problem like the health of our planet. What they all agree on is that harm is taking place and we have to change things.

2
WHAT STOPS PEOPLE

"The only way forward, if we are going to improve the quality of the environment, is to get everybody involved."

— RICHARD ROGERS

FROM YEARS OF TRYING to teach people and companies to make sure they obeyed the environmental rules, recycle, conserve and practice sustainability I have seen many things contribute to the "why" people don't do many of the things we should. Excuses like just it is just not the easy way or we don't have enough space where we live, going for the cheapest item, saying if they paid me to recycle I would, using the strongest cleaner instead of the safest, or just not having the time, and my favorite is that it is just too hard. One last reason in this list is that some people just have no desire and believe the earth is here to serve them, now how arrogant and uniformed of the facts is that one. The biggest argument often is it is just too expensive. Well, the reality is none of the above is true with one exception which is it does take some time to learn yourself and those close to you about how to do things that can make a difference, and yes, some can be fun and save money but make you some in the short and long run but all will have an impact on saving this planet.

3
WHY WE SHOULD

"The greatest threat to our planet is the belief that someone else will save it."

— ROBERT SWAN, AUTHOR

WHY YOU SHOULD do some of this stuff may not be understood just yet depending on your age and how much you have read, or seen on TV or adults have shared with you. I promise that if you do some of the activities below and look at some of the recommended videos besides what is mentioned above you will begin to understand why this is so important. So, if you can only do a few things offered please involve your friends and ask adults in your life to do the activities with you or help you get started. To start you on your journey I suggest you look at the following videos, one or more below should help you understand why recycling and conservation is so important. The first 7 are for the younger kids from 4-11 and 8 and 9 are for the pre-teens and teenagers, but as always parents should take a look at the ones for the kids first and even the ones for the pre-teens and teenagers as the age group appropriateness is only my view and your view may differ.

1. https://www.youtube.com/watch?v=vNyv4fGRO5o
   Why Is Recycling So Important / Kids Video Learn why recycling is so important with Natalie and Matt!
2. https://www.youtube.com/watch?v=BaFpv03hq-4
   Uploaded by WonderGrove Kids Recycling It's my turn to

help.
3. https://www.youtube.com/watch?v=BaFpv03hq-4
   The 3 R's for Kids
4. https://www.youtube.com/watch?v=OasbYWF4_S8
   Reduce, Reuse and Recycle, to enjoy a better life
5. https://www.youtube.com/watch?v=OD-Xp1BU874
   Rabbit Sings A Song About Recycling
6. https://www.youtube.com/watch?v=rl0YiZjTqpw
   Save Water to Help the Earth
7. https://www.youtube.com/watch?v=PqxMzKLYrZ4
   Global Warming for Kids
8. https://www.greenbiz.com/article/12-eye-opening-sustainability-videos-2016

Sustainability has been complicated as the world global temperatures rise and the site above from GreenBiz addresses the issue impactful. The site offers 12 videos that tell a sustainability story clearly and creatively in 15 minutes or less. Two of them at a minimum I suggest you view are The Refuge and the Deforestation Film.

9. https://www.youtube.com/watch?v=qBBOue_AdcU&index=5&list=PL5WqtuU6JrnXjsGO4WUpJuSVmlDcEgEYb

Below is a listing Conservation International's well done and poignant "Nature is Speaking" video series that uses dramatic celebrity voice-overs to where the celebrities represent different aspects of our environment (anthropomorphize) that are the powerful natural forces we take for granted. It warns us that we the Earth Dwellers need nature; NATURE doesn't need the Earth Dwellers. The videos are all great but I have listed below the ones I would like you view at a minimum.

   1. Nature Is Speaking: Joan Chen is Sky | Conservation International (CI)

2. Nature Is Speaking: Shailene Woodley is Forest
3. Nature Is Speaking: Lee Pace is Mountain | Conservation International (CI)
4. Nature Is Speaking: Reese Witherspoon is Home | Conservation International (CI)
5. Nature Is Speaking: Liam Neeson is Ice | Conservation International (CI)
6. Nature Is Speaking: Julia Roberts is Mother Nature | Conservation International (CI)

Recycling Facts Review: This second part of this chapter is not an activity but information sharing. A lot of information is provided below and I wanted to provide a general overview on the subject of recycling but wanted you to trust the source so all the information I am providing is a copulation of information from very trusted sources, in my view and experience. A lot of sources are out on the internet and many federal, state, county, city, non-profit organizations, and educational institutions provide a vast amount of information and guidance about recycling which is readily available if you really want to dig deeper. You may think my sources are limited and you can find contradictory facts provided from various organizations and the more you get involved and learn you can determine for yourself what makes sense. I am not saying that I personally agree with all the facts provided in this section but this is not the platform to debate certain perspectives or data provided. As you learn about all the issues, which I really hope you pursue more knowledge about, don't hesitate to challenge others respectfully and with the facts and not just some feeling that a position or fact is wrong. It is my life experience in this field that most people are sincere in their beliefs as a quote from Mark McKinnon who is an American political advisor, reform advocate, media columnist and television producer states "Reasonable people can reasonably disagree".
References For This Section:

Environmental Protection Agency (EPA): Available at https://www.epa.gov/recycle/recycling-basics and https://www.epa.gov/sites/production/files/2015-11/documents/sfhomes.pdf

Massachusetts Institute of Technology (MIT): Available at http://web.mit.edu/facilities/environmental/recyc-facts.html

National Resource Defense Council (NRDC): The NRDC combines the power of more than three million members and online activists with the expertise of some 500 lawyers, scientists, and policy advocates to secure the rights of all people to clean air, clean water, and healthy communities and have been doing their work 1970, with a powerful track record of success. Available at http://www.nrdc.org/cities/recycling/gsteps.asp

Keep America Beautiful (KAB): The KAB has a network of more than 600 state and community-based affiliates that carry out our shared mission at the state, county and municipal levels. Available at http://www.kab.org/site/PageServer?pagename=recycling_facts_and_stats

Texas Take Care of Texas: The Take Care of Texas program is a statewide campaign from the Texas Commission on Environmental Quality that provides helpful information on Texas' successes in environmental protection and encourages all Texans to help keep our air and water clean, conserve water and energy, reduce waste, and save a little money in the process! Available at http://takecareoftexas.org/kids/fun-facts. Please note that this is included because it is my state and I am very familiar with their efforts and perspective but please check out your state, county, or city programs because there are a lot of them out there that do a tremendous amount of great work.

Eco-Cycle (EC): EC is one of the largest non-profit recyclers in the USA and has an international reputation as a pioneer and innovator in resource conservation. Eco-Cycle's mission is to identify, explore and demonstrate the emerging frontiers of sustainable resource management through the concepts and practices of Zero Waste. They believe in individual and community action to transform society's throw-away ethic into environmentally-responsible stewardship. Available at https://www.ecocycle.org

The statements from each organization are identified by the organization abbreviation at the end of each statement.

What Does It Mean To "Recycle":
Recycling is the process of collecting and processing materials that would otherwise be thrown away as trash and turning them into new products. (EPA)

Waste Related Things We Do:
It is currently estimated that we are consuming at a rate of 123%, using 23% more resources than the Earth can sustain. This means we are jeopardizing the ability of future generations to meet their needs. The effects of this over-consumption can also be seen today in everything from acid mine drainage and declining biodiversity to increasing conflict over access and ownership of the shrinking natural resources on the planet. (EC)

Our current system of one-way use is not just. As populations continue to increase and seek the affluence and consumerism of the Western culture, there will be increasing conflict over our limited supply of resources, everything from precious metals to clean water. Over the past couple of years, China's rapid growth has driven the price of metal to all-time records, levels high enough to entice criminals to steal manhole covers, aluminum siding from homes, and copper wire and pipes from all over the U.S. There are violent, ongoing conflicts in Africa over rights to precious minerals such as the diamond wars in Sierra Leone, Liberia, and Guinea, and a coltan (mineral used in cell phones and computer chips) war in the Democratic Republic of the Congo. Wars for fossil fuels and clean water are already developing. (EC)

In 2015, Americans recycled or composted about 1/3 of their waste. (TCOT)

The average Texan throws away about 6 and a half pounds of trash each day. (TCOT)

The average person generates over 4 pounds of trash every day and about 1.5 tons of solid waste per year. (EPA)
In 2009, Americans produced enough trash to circle the Earth 24 times. (EPA)

Over 75% of waste is recyclable, but we only recycle about 30% of it. (EPA)

We generate 21.5 million tons of food waste each year. If we composted that food, it would reduce the same amount of greenhouse gas as taking 2 million cars off the road. (EPA)

Allow a faucet leaking will leak at a rate of one drop per second can waste up to 3,000 gallons of water per year. That's the amount of water needed to take more than 180 showers! (TCOT)

In the past 50 years, humans have consumed more resources than in all previous history. U.S. EPA, 2009. Sustainable Materials Management: The Road Ahead. (EC)

Americans throw away 25,000,000 plastic bottles every hour. (EPA)

The average American throws away 3.5 pounds of trash per day. (MIT)

The average American uses 650 lbs. of paper per year and one ton of paper from recycled pulp saves 17 trees, 3 cubic yards of landfill space, 7000 gallons of water, 4200 kWh (enough to heat a home for half a year), 390 gallons of oil, and prevents 60 pounds of air pollutants.(MIT)

In the past 50 years, humans have consumed more resources than in all previous history. U.S. EPA, 2009. Sustainable Materials Management: The Road Ahead. (EC)

Between 1950 and 2005, worldwide metals production grew six fold, oil consumption eightfold, and natural gas consumption 14-fold. In total, 60 billion tons of resources are now extracted annually — about 50% more than just 30 years ago. Today the average European uses 43 kilograms of resources daily, and the average American uses 88 kilograms. Worldwatch Institute, 2010. State of the World 2010. (EC)

Between 1970 and 1995, the U.S. represented about one-third of the world's total material consumption. With less than 5% of the world's population, the U.S. consumes 33% of paper, 25% of oil, 15% of coal, 17% of aluminum, and 15% of copper. U.S. EPA, 2009. Sustainable Materials Management: The Road Ahead. (EC)

More than 100 billion pieces of junk mail are delivered in the United States each year, which comes out to 848 pieces per household. The production, distribution and disposal of all that junk mail creates over 51 million metric tons of greenhouses gases annually, the equivalent emissions of more than 9.3 million cars. ForestEthics, 2008. Climate Change Enclosed: Junk Mail's Impact on Global Warming. (EC)

The U.S. buried or burned more than 166 million tons of resources—paper, plastic, metals, glass and organic materials—in landfills and incinerators in 2008. We recycled and composted only one-third of our discards. U.S. EPA, 2009. Municipal Solid Waste Generation, Recycling, and Disposal in the United States, Detailed Tables and Figures for 2008. (EC)

We have destroyed half the world's tropical and temperate forests and they are now gone. U.S. EPA, 2009. Sustainable Materials Management: The Road Ahead. (EC)

We have used up 75% of marine fisheries that are now overfished or fished to capacity. U.S. EPA, 2009. Sustainable Materials Management: The Road Ahead.

Why Recycle:

Recycling is one of the most feel-good and useful environmental practices around. The benefits go way beyond reducing piles of garbage -- recycling protects habitat and biodiversity, and saves energy, water, and resources such as trees and metal ores. Recycling also cuts global warming pollution from manufacturing, landfilling and incinerating. (NRDC)

Helps sustain the environment for future generations; (EPA)

Helps create new well-paying jobs in the recycling and manufacturing industries in the United States. (EPA)

It saves energy, preserves natural resources and wildlife, and reduces our carbon footprint. (NRDC)

Recycling waste materials is better than landfilling them, but the best option is to not generate them at all. (NRDC)

Conserves natural resources such as timber, water, and minerals; (EPA)

Prevents pollution by reducing the need to collect new raw materials, saves energy, and reduces greenhouse gas emissions that contribute to global climate change; (EPA)

Water is a finite resource — even though about 71 percent of the Earth's surface is covered by water, less than one percent is available for human use. (TCOT)

Aluminum:

Recycling one aluminum can saves enough energy to listen to a full album on your iPod. Recycling 100 cans could light your bedroom for two whole weeks. (EPA)

Recycling aluminum cans saves 95% of the energy used to make alum cans from new material. (EPA)

One recycled aluminum can will save enough energy to run a laptop computer for over five hours. (TCOT)

Recycling a soda can saves 96 percent of the energy used to make a can from ore and produces 95 percent less air pollution and 97 percent less water pollution.(MIT)

Recycling a soda can saves 96 percent of the energy used to make a can from ore and produces 95 percent less air pollution and 97 percent less water pollution.(MIT)

It only takes about 6 weeks total to manufacture, fill, sell, recycle, and then remanufacture an aluminum beverage can. U.S. EPA, 2010. Common Wastes & Materials: Aluminum. (RC)

Paper:

One ton of paper manufactured from recycled paper saves up to 17 trees and uses 50 percent less water. (TCOT)

Every ton of newspaper recycled saves 4100 kWh or enough energy to power a TV for 31 hours. (MIT)

Every ton of newspaper recycled saves 4100 kWh or enough energy to power a TV for 31 hours. (MIT)

Oil:

The oil from one oil change released into the environment can contaminate up to one million gallons of drinking water—a year's supply for 50 people. (TCOT)

One gallon of oil, when reprocessed, can generate enough energy to meet the electricity needs of a home for half a day (MIT)

Composting:

In 1980, recycling and composting kept 14.5 million tons of trash from landfills and incinerators in the United States. By 2013, that number had reached 87.2 million tons that accounts for only 34.3 percent of all the stuff we discard in the USA every year. (NRDC)

Things To Do To Reduce Waste:

Reduce waste by buying in bulk, using reusable shopping bags and water bottles, and tossing out less food. (NRDC)

Reuse clothing and other unwanted items by donating them to community organizations like homeless shelters, thrift stores, and animal shelters. (NRDC)

Rethink what you typically throw out—like grass clippings, which can be left on your lawn to condition and fertilize the soil. Cutting down on waste could even save you money, since a growing number of communities have "pay as you throw" programs that charge a waste collection fee based on the size of your garbage can. (NRDC)

Paper:

Making paper from recycled fibers, for example, uses less energy and water and produces less air and water pollution than making paper from trees. (NRDC)

Aluminum:

Used aluminum cans are recycled and back on the shelf as new cans in as few as 60 days. Twenty recycled cans can be made with the energy needed to produce one can using virgin ore. Recycling one aluminum can saves enough energy to run your television for three hours. The amount of energy saved just from recycling cans in 2010 is equal to the energy equivalent of 17 million barrels of crude oil, or nearly two days of all U.S. oil imports. (KAB)

The pollutants created in producing one ton of aluminum include 3,290 pounds of red mud, 2,900 pounds of carbon dioxide (a greenhouse gas), and 81 pounds of air pollutants and 789 pounds of solid wastes. Tossing away an aluminum can wastes as much energy as pouring out half of that can's volume of gasoline. (KAB)

Glass Bottles:

Recycling one glass bottle saves enough energy to light a 100-watt light bulb for four hours, power a computer for 30 minutes, or a television for 20 minutes. (KAB)

Education, Knowledge, Things We Need To Do:

"Only when the last tree has been cut down Only when the last river has been poisoned Only when the last fish has been caught Only then will you find that money cannot be eaten."
- Cree Indian Prophecy

Develop systems that reflect a "Zero Waste" approach that refers to redesigning of our production and consumption systems to use resources more efficiently, to prevent waste before it happens, and to incorporate all leftover materials back into the production cycle rather than discarding them as waste. If we pair Zero Emissions and Zero Waste together, we have our solution. Each of us has an important role to play to make this happen. (EC)

Most environmental impacts associated with the products we buy occur before we open the package, so buying products made from recycled materials is just as important as sorting waste into the right bins. And when we reduce the amount of stuff we buy in the first place, and reuse what we can, we reduce the environmental harm associated with acquiring raw materials and manufacturing. (NRDC)

"The power of a lot of people acting correctly is the most important thing," says Darby Hoover, an NRDC senior resource specialist. "People need to familiarize themselves with community guidelines. Visit your municipality's website to learn more about your local recycling rules and options and don't forget to check whether your community collects food or yard waste for composting, as well. (NRDC)

Contact your local government's recycling or solid waste department to learn more about the services it provides. Your current waste haulers may offer information as well. Also consider joining the EPA's free WasteWise program, which provides members with several benefits, including a technical assistance team that can help you conduct a waste audit, reduce waste, and implement a recycling program. (NRDC)

Buy recycled to make sure the products you purchase (and the packaging they come in) are recyclable, and when possible, choose products that contain recycled materials. "If you purchase products that have recycled content, you're closing the recycling loop and making sure the cycle continues. (NRDC)

Raise community awareness just like Mr. Hoover of the NRDC states "Schools are great places to educate about recycling," If you are a parent, teacher, or student, get the other members of your community involved in an informal waste audit: Take a look in a school trash bin and ask what's there and how and if it can be recycled or composted. Plan a community tour of your local materials recovery facility, recycling center, or landfill. (NRDC)

Support legislation that holds manufacturers financially accountable, you encourage the design of better and more recyclable goods and ensure that producers pay their fair share of the costs of recycling by being an extended producer. An Extended producer responsibility (EPR) is a form of product stewardship. Under EPR, manufacturers and brand owners (known as producers) are responsible for the products they make or sell, and any associated packaging, when they become waste. . (NRDC)

For every one million tons of material recycled rather than landfilled, we save the energy equivalent of Aluminum: 35,680,000 barrels of oil, Glass: 460,000 barrels of oil, Newspaper: 2,920,000 barrels of oil, Office paper: 1,760,000 barrels of oil, Mixed residential paper: 4,010,000 barrels of oil, PET (plastic): 9,100,000 barrels of oil, and HDPE (plastic): 8,870,000 barrels of oil (EC)

What To Do Next:

"We must become the change we want to see."
- Mahatma Gandhi

Now that you have some facts, besides doing some of the activities in the rest of this book, hopefully you will want to explore more about one of the facts or comments provided or maybe even use them as a basis for a school paper but most of all I hope that it helps you to understand how important recycling is to saving this planet and all of us can do our part, no matter how large or small to help make a change as Earth Dwellers that will improve our environment now and for future generations.

4
STARTING OFF WITH THE THINGS FOR TO DO AT HOME AND ALL THE FAMILY CAN PARTICIPATE IN

PLASTIC OR PAPER OR CLOTH:

Kids Age Range – All Can Participate

The "bag" issue has always been a significant problem and has been driven a great deal by product cost to the store and industry capability to manufacture them. I have included a significant amount of information on this issue below for two reasons 1) to explain the significance of the problem, and 2) in hopes that it will motivate you to take a personal step with your family to help deal with this catastrophic issue. If you don't do a single other thing recommended by this book but eliminated plastic bags in your life where possible it will be a victory for this Earth and all who live on this planet. The plastic bag abundance has not been the problem it is today until after the late 1970s, single-use plastic bags were seldom available in grocery stores. Since then an estimated one trillion (1000 billion) bags are used each year globally, but they are so seamlessly ingrained into our daily routines that we hardly notice. It is difficult to imagine life without the plastic bag. The average American throws away about 10 single-use plastic bags per week. While their environmental costs are really a problem for communities and the planet, the cost of plastic bags for stores is pretty low. Made from ethylene, a byproduct of petroleum or natural gas, plastic bags are so cheap and flimsy that cashiers use them freely, double bagging as a matter of course and often sticking just a few items in each bag. As a result, shoppers end up with piles of plastic bags, I know some of us reuse plastic shopping bags to line our waste bins or to pick up dog poop, but the bags still end up in the landfill. New Yorkers alone use twice the national average. Some 23 billion are used across the state each year, more than enough, when tied together, to stretch to the moon and back 13 times. The United Nations Environment Program estimates that some eight million tons of plastic waste that includes plastic bags end up in the oceans each year, while a 2016 scary report by the World Economic Forum report projects that there will be more plastic than fish by weight in the oceans by 2050 if current trends continue. At the same time the plastic production and disposal also generates around 400 million tons of carbon dioxide that hurts the atmosphere and impacts the air we breathe and the whole environment, a year globally.

If you fish, hunt or just enjoy the outdoors and nature you should know that millions of whales, birds, seals and turtles have been killed because they mistake plastic bags for food or because they become entangled in packing bands (the plastic holding cans together that I explain more about later) and other items. Trillions of micro plastics end up in the ocean, with seafood eaters ingesting an estimated 11,000 tiny pieces annually. If you think it is just nature's problem you should know that plastic fibers have also been found in tap water around the world; in one study, researchers found that in 94 percent of water samples in the United States.

Plastic bags present a number of additional problems besides what is mentioned above along with more facts as to their harm:

1. They are made from fossil fuels
2. A source of litter on land and in waterways
3. A source of avoidable excess packaging waste used for mere minutes
4. Problematic, creating tangles and jams in recycling and waste water processing equipment.

5. Costly for municipalities and recycling centers in terms of time and money to recycle and in many cases, municipalities have stopped taking plastic.
6. Americans use 100 billion plastic bags a year which require 12 million barrels of oil to manufacture. So it not just the pollution and harm to humans, wildlife and our waters it requires a lot of oil. It only takes about 14 plastic bags for the equivalent of the gas required to drive one mile.
7. The average American family takes home almost 1,500 plastic shopping bags a year. You may think this is a lot but I encourage to do the math for your family, don't forget to count all sources not just your weekly or biweekly grocery store trip.
8. According to Waste Management, only 1 percent of plastic bags are returned for recycling, that means that the average family only recycles 15 bags a year; the rest ends up in landfills as litter. This is sad but when you also add that municipalities have stopped allowing them because of the "problems". What do they expect us to do with them eat them? Our government is supposed to help protect our environment, here many have failed with the exception of California and a few others that have banned them in many areas.
9. Up to 80 percent of ocean plastic pollution enters the ocean from land.
10. At least 267 different species have been affected by plastic pollution in the ocean.
11. 100,000 marine animals are killed by plastic bags annually is an estimate only based on a lot of field work.

From Wallace "J." Nichols, a marine biologist and research associate for the California Academy of Sciences identifies that while a lot of figures have been thrown around in the media, hard numbers are very difficult to calculate, and the sad fact is that when most sea animals eat plastic and die, they sink to the bottom, unaccounted for. But possibly more significant than the individual animals that are killed by eating plastic are those that are affected indirectly. For example, when sea turtles eat plastic instead of food, their glucose levels drop, leaving them with less energy for migration and reproduction. Females can't lay as many eggs, and fewer new sea turtles are born. "When you connect the dots," J." said, "you realize that plastic pollution may cost millions of potential sea turtle lives."

12. One in three leatherback sea turtles have been found with plastic in their stomachs.
13. Plastic bags are used for an average of 12 minutes is a sad statistic but think about all the bags that gather in your pantry if like me that is where you store the ones you can't avoid even though we try to re-use them at some point they are put into the recycle bin. We are lucky in our area they do take them in the recycle but they still stay buried someplace.
14. It takes 500 (or more) years for a plastic bag to degrade in a landfill. This is worse case as many will only take 5- 10 years to decompose but either way it is a long time. Also, decomposition does not mean the chemicals go away just that they are not in bag form anymore. Unfortunately, the bags don't break down completely but instead photo-degrade, becoming micro plastics that absorb toxins and continue to pollute the environment.
15. It is not just a USA problem, a Google search on "animals eat plastic bags" brings up hundreds of heartbreaking stories and images from around the world. Large numbers of foraging cows in India have died from ingesting plastic bag litter even though in that country they have banned the distribution of plastic bags. In the United Arab Emirates, a veterinarian has documented images of camels,

sheep, goats, and endangered desert animals that die from eating plastic bags. Whales wash up on our coasts, their bellies full of plastic.

Okay it is not just a problem impacting water and aquatic life but Plastic affects human health. Toxic chemicals leach out of plastic and are found in the blood and tissue of nearly all of us. Exposure to them is linked to cancers, birth defects, impaired immunity, endocrine disruption and other ailments. Below are two videos I strongly recommend, one for adults and one for the kids. The first titled "Peril of Plastics: risks to human health and the environment" is from a professor at Arizona State University (2010) and identifies issue of concern that is worth your time.
https://biodesign.asu.edu/news/perils-plastics-risks-human-health- and-environment

The second is a PBS video from YouTube as part of the Nature Cat Series that addresses the plastic bags at a kids entertaining level. It is for younger kids and would be great if older kids and or adults watched it with them as it would be great time to talk with them about assisting in the plastic bag conversion to paper or hopefully cloth.
https://www.youtube.com/watch?v=Wi99ypudQ3c

So to start your conversion to a non-plastic bag family you need to have your cloth / cotton bags available and below is a pictures of typical bags available at most places like Walmart, Target, Cosco, and of course on line many others are available. Get at least a dozen to have back-ups and in both your cars if you are a two car family. If you do get plastic bags for some reason you should try to re-use them as often as possible and recycle if you can. You can also use your own paper bags if that is your preference and get strong ones so they can be re-used several times. If you have to use paper and dispose of it at least it won't be in the landfill for 500 years as some plastics will be and will biodegrade within just a few weeks.

As a side note, if you or the adults in your life buy in bulk you have less plastic in the packaging per serving and it can go a long way in reducing the plastic you generate.

An example of Cotton Bags found on Amazon from Simple Ecology at a cost on 09/01/18 of $21.95

Designate A Container In Your Room For Scrap Paper:

This activity can really do a lot of good over time because it actually can save trees. You can really save trees on your own not just what efforts you may do at school. You just need to set up a container in your room or even in the kitchen where you and everybody in the house puts paper that you only write on one side of. No matter how big or small it all counts so it is a good way to get the whole family thinking recycle. The process is pretty simple and as follows:

1. Put a trash can in a convenient spot either in your room or in the kitchen and clearly mark it RECYCLE PAPER.
2. Have a discussion with the rest of the family and explain to them what you want to do and why. WHAT you are asking to collect all sizes of any color paper that has only one side written on. The WHY is rather straight forward as recycling 1 ton of paper can save 17 trees, 7,000 gallons of water, 2 barrels of oil, and 4,000 kilowatts of electricity. The energy that you save can power 1 home for 5 months. The most important point for the family is that the average family uses 6 trees worth of paper each year. So if you could just reduce the amount of paper by half you could actually save 3 trees which would be great. Per the United States Environmental Protection Agency (EPA) "If every American recycled just one-tenth of their newspapers, we could save about 25 million trees each year." It would be so great to get lots of families doing this just a little bit and we could easily save millions of trees really helping this planet and all the people and living things on it.
3. With all the paper collected and your container full your work begins because you have promised to be the one to take care charge of the effort.

A. You need to sort the paper by size.
B. Draw a big x on the side that has writing.
C. Staple same sizes of paper together in 10 piece pads for reuse. Two staples at top one on the right and one on the left should be plenty and for small pads you can use just one staple in the left top corner.
D. If not enough to have a pad of one size then cut up to make small note pads.
E. You will be surprised how many pads you may come up with and they can be used for notes and lists for all sorts of things like groceries, phone call messages, family notes that you post on the refrigerator like a lot of us do.
F. When they are used for the second time then they go into the recycle bin that I hope you already have, if not then you should talk to your parents or guardian and find out why not and hopefully start using one that they will need to help you get.

I promise if you do this you will be surprised on how much paper can really be reused and the trees you save will really do some good for this planet.

PLASTIC RINGS ON DRINK CANS:

(Six pack rings or six pack yokes are a set of connected plastic rings that are used in multi-packs of beverage, particularly six packs of beverage cans.)

Kids Age Range – All Can Participate

I have always wondered as to why people don't understand how harmful plastic rings on drink cans are to birds and even seals and turtles. Even though many are photodegradable (which means they break down in the sun) six-pack rings do break down when exposed to summer sunlight, a photodegradable plastic ring carrier will begin to lose its strength in a short period of time and become totally brittle in about three to four weeks.

However, if not recycled or disposed of properly they can be brutal on many species.

The 6-pack rings impact thousands of birds, turtles, marine mammals, and other wildlife who are killed every year by discarded 6-pack rings. Some wildlife get entangled in the rings as they wrap around their beak or muzzle, preventing them from eating. It tangles up their feet, wings, or fins. Young wildlife get entrapped and as they grow the 6-pack cuts into their flesh, sometimes cutting off limbs or killing the animal.

Other animals mistake the floating object as food and eat it. Think about it for a second before you judge them as stupid creatures. Don't start thinking it is the smart wildlife world and the smart ones as a group will be helped if the stupid ones are taken out because plastic floating on the surface of the water can resemble many edible items? It can look like a sea jelly (a.k.a. jellyfish) or a piece of seaweed. Since plastics have only been around since the late 1970s so most critters haven't got trained by other generations that it shouldn't be eaten. On the side of fairness 6-pack rings are not a major contributor to environmental pollution and wildlife mortality. Other plastic products are far more abundant and deadlier. Cigarette butts, monofilament fishing line, and disposable plastic lighters are bigger dangers than 6- pack rings. But it all counts and this is one that is easily managed for benefit of all.

Some difference of opinion on this issue is out there because 6- pack rings are now required to be photodegradable they say you should not need to cut them up, oh really, then who watches them from the time it is thrown away and it breaks down so the birds, fish, and other animals don't suffer from your unwillingness to cut them up.

Here is a YouTube video on how to do it if you need pictures: https://www.youtube.com/watch?v=PPaU6BEP3uo

Photos below are really sad and show the impact of 6-pack rings discarded and the poor unfortunate creatures that found your discard.

Photo by Stephane Oravetz from Article BY ELIZABETH CLAIRE ALBERTS

Humane society of the United States

Photo Credit: www.forums.miamibeach411.com

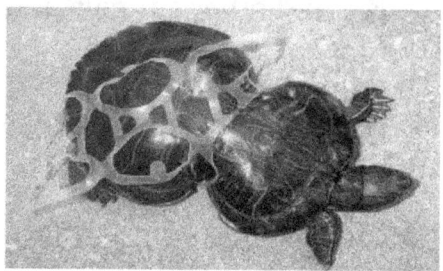

National Geographic Education Blog 05/20/2016

With the problem comes a quick and easy solution that as a kid you can do if you are old enough to handle scissors or adults can help you do it. From the Fox Hollow Cottage at right is a simple reminder and approach to the problem by cutting all the circles and other holes that make up the ring.

ALUMINUM CANS:

Kids Age Range – All Can Participate

To start this one off I want to suggest you go to a video from Chula Vista Clean & Green Curiosity Quest https://www.youtube.com/watch?v=jHx95PQIl4k to view how easy it is and that you can actually make a few bucks. Motivation for you as a kid may be that if you do it all by yourself or with your family or friends you will be able to get some of the money for recycling. It is a great activity for all involved.

Aluminum recycling benefits the environment and future generations by saving an estimated 95% of the energy required for aluminum production from ore, thus greatly reducing air emissions including greenhouse gases. Alumina, the raw material for primary aluminum production, is extracted from bauxite. For every pound of aluminum recycled, four pounds of bauxite are conserved. Over 50 million pounds of aluminum cans are recycled every week. All recycling efforts save critical space in landfills.
Because so many of them are recycled, aluminum cans account for less than 1% of the total U.S. waste stream, according to EPA estimates. Americans throw away enough aluminum every month to rebuild our entire commercial air fleet. A steel mill using recycled scrap reduces water pollution, air pollution, and mining waste by about 70 percent.

On average, Americans drink one beverage from an aluminum can every day. But we recycle just over 49% of the cans we use.

An aluminum can that is thrown away will still be an aluminum can 500 years from now!

WHAT NEEDS TO BE DONE?

1. Collect the cans in a separate clean container or plastic bag and make sure the can is empty prior to storage. You don't have to rinse it but make sure it is as empty as possible because you don't want to attract bugs while it is stored.
2. I suggest you periodically squash the cans, to get more in your container and less frequent trips to the recycler.
    a. Look at the can flattening as fun for the kids but be safe. There are also manual helpers that work well and electric ones too.
3. You can find a recycler in your area on the web or even call your state agency and they can tell you where the closest recycler is.
4. Gather up the cans, collect the kids and if needed get an adult to drive so you all can see the end of the process and collect your money.
    a. Currently, eleven states and Guam in the U.S. have active bottle bills which require minimum payouts of between 5 cents and 10 cents per recycled aluminum can. This means that if you live in one of the ten states with bottle bills, your income potential for recycling aluminum cans just went up astronomically.
    b. For none bottle bill states prices go up and down but are usually similar throughout the country at any given time. With approximately a half-ounce of aluminum per can, or 29 cans per pound. With a $0.60 per pound market rate for general aluminum, the least you should receive for your cans is $0.30 per pound. Anything less than that is a scrap yard dealer attempting to maximize their profits at your expense.
    c. Think about it, it won't take long to collect a few hundred cans, average households generate about 350 cans per person per year. So a family of 4 could generate at least 1400 cans per year not counting parties or group gatherings at home that can generate more cans. Conservatively your family could generate 1800 cans per

year at a going rate of $0.60 per pound and 1800/29 gives you approximately 62 pounds times $0.60 = $37.00 at a minimum. But, if you live in one of the ten states that have "bottle" bills you could make $90 to $180 a year which is not bad at all.

d. Oregon was the first state in the union to pass a "bottle bill", in 1971. As of October-2010 there are 11 states and Guam that have container deposit laws:
  i. California,
  ii. Connecticut,
  iii. Delaware,
  iv. Hawaii,
  v. Iowa,
  vi. Maine,
  vii. Massachusetts,
  viii. Michigan,
  ix. New York,
  x. Oregon,
  xi. Vermont.

Take that money and spend it wisely or maybe buy something you really wanted and makes it worthwhile for all the effort while you help save the planet.

USE A PLASTIC DRINKING BOTTLE = NOT DISPOSABLE:

Kids Age Range – All Can Participate

- In the U.S., 1,500 plastic water bottles are consumed every second.
- China, USA, Mexico, and Indonesia27 are the four largest consumers of bottled water.

From "One Green Planet" some information on plastic bottles containing Bisphenol A (BPA).28 which is a chemical found in many hard plastics that we use every day. Higher doses have been linked to infertility and other health problems. However a recent (2018) FDA study identified in a report released by the National Toxicology Program indicates toxicity is not as bad as many advocacy groups have indicated and levels you are exposed to are not harmful. As in many issues people disagree on the amount of harm and this big study has yet to have all the review needed by other scientists to be totally validated and I would also argue that with the limited time frame of human exposure overall we just don't really have the absolute answer yet. So you can understand the possible negative side I have information below. Because it is better to be safe than sorry your avoidance of the exposure can only be positive besides the other big reason in how much you not using all those plastic bottles will help the environment. I don't want to scare you but inform you about stuff you should be aware of and facts that may actually make you think about going to a hard plastic bottle (sports bottle) and use filtered water from your own home, yes water at home can taste good, instead of those plastic supposedly recyclable bottles. One qualifier is that the information applies to most "plastic" disposable bottles and you should actually verify what you are using because of the possible exposure issue identified below. Use a BPA-free reusable bottle (explained below). Carry a refillable, BPA-free bottle when you are on the go, and refill it whenever the option arises.

The Human Impact: Plastic bottles contain Bisphenol A (BPA), the chemical used to make the plastic hard and clear. BPA is an endocrine disruptor which has been proven to be hazardous to human health. It has been strongly linked to a host of health problems including certain types of cancer, neurological difficulties, early puberty in girls, and reduced fertility in women, premature labor, and defects in newborn babies, to name a few examples. BPA enters the human body through expo- sure to plastics such as bottled drinks and cleaning products. It has been found in significant amounts in at risk groups such as pregnant women's placentas and growing fetuses. A study conducted last year found that 96% of women in the US have BPA in their bodies. The good news is that you can have your BPA levels measured and make lifestyle changes to lower them, as demonstrated by Jeb Berrier in his film about plastic consumer merchandise balled "Bag It". (Go to http://www.bagitmovie.com/about.html for more information on the movie) Bottled drinks also contain phthalates, which are commonly used in the U.S. to make plastics such as polyvinyl chloride (PVC) more flexible. Phthalates are also endocrine-disrupting chemicals that have been linked to a wide range of developmental and reproductive effects, including reduced sperm count, testicular abnormality and tumors, and gender development issues. The FDA does not regulate phthalates or class them as a health hazard due to the supposedly minute amounts present in plastic bottles. This decision does not take into account the significant presence of plastics in the average American citizen's daily life, the fact that phthalate concentration increases the longer a plastic water bottle is stored, or that a bottled drink that is exposed to heat causes accelerated leaching of harmful plastic chemicals into the drink.

In addition to the negative impacts of BPA and phthalates on human health there are also growing concerns regarding carcinogens and microbial contaminants that have been found in test samples of bottled water.

The Animal Impact: Plastic bottle tops are rarely recyclable as with plastic bags they often end up at the bottom of the ocean, and in the stomachs of a variety of animal species that mistake them for food. One albatross that was recently found dead on a Hawaiian island had a stomach full of 119 bottle caps. Marine life falls prey to this problem on a daily basis. A sperm whale was found dead on a North American beach recently with a plastic gallon bottle which had gummed up its small intestine. The animal's body was full of plastic material including other plastic bottles, bottle caps and plastic bags.

The Environmental Impact: Plastic bottles are made from a petroleum product known as polyethylene terephthalate (PET), and they require huge amounts of fossil fuels to both make and transport them. In the 1970s the U.S. was the world's largest exporter of oil, but now it is the largest importer next to China. If you fill a plastic bottle with liquid so that it is 25% full, that's roughly how much oil it took to make the bottle. For a single- use disposable item, that's a lot.

It's harder to recycle plastic bottles than you think. Of the mass numbers of plastic bottles consumed throughout the world, most of them are not recycled34 because only certain types of plastic bottles can be recycled by certain municipalities. These plastic bottles that are not recycled either end up lying stagnant in landfills, leaching dangerous chemicals into the ground, or they infiltrate our streets as litter. They are found on sidewalks, in parks, front yards and rivers, and even if you chop them into tiny pieces they still take more than a human lifetime to decompose.

It gets worse, in the case of bottled water, the plastic-making process requires over two gallons of water for the purification process of every gallon of water. In the U.S., bottled water and tap water are regulated by different federal agencies. The Food and Drug Administration (FDA) regulates bottled water and the Environmental Protection Agency (EPA) regulates tap water. Therefore, the enforcement and monitoring of water quality for bottled water vs tap water does not add up. Due to strict EPA policies, incidents of tap water contamination has to be reported immediately to U.S. citizens, however there is no such rule for bottled water, despite numerous bottled water recalls 35 taking place over the years.

Although several scientific studies have been done concerning the problems of chemicals found in bottled drinks, there have been various campaigns to undermine the results of the research. The American Chemical Council (ACC) still claims that BPA is safe." From an energy consumption perspective, you should know that manufacturing a quart (liter) plastic bottle requires 2.6 oz. (100ml) of crude oil. Plastic drink bottles are made from a polymer called PET, which stands for Polyethylene Terephthalate. The chemicals used to make PET, xylene and ethylene, are extracted from crude oil. The other main ingredient in PET manufacturing is natural gas. Making a quart bottle requires 1.5 cubic feet (42 liters) of natural gas. That's a sphere 53 inches (135 cm) wide. If you do the math for 1500 bottles a second, we are using approximately one barrel (42 gal) of crude per second, 60 barrels a minute, 3,600 barrels an hour, and 86,400 barrels a day just for plastic bottles that cause so much harm. This doesn't count the natural gas, and fuel involved in the manufacturing process and transportation to get them to you.

So get those BPA free bottles for home, the car, and to take one while you are walking, jogging or working out or just sitting around and start saving the planet.

Pack A Waste Free Lunch:

Kids Age Range – All Can Participate, But Will Probably Need Adults Help But A Great Family Activity

Packing a Waste-Free Lunch can be a little challenging when you first start but can be pretty routine once you get all the reusable stuff together and you do so much good for the environment in eliminating plastics and disposable containers from the environment. The whole objective for this activity is to try to eliminate as much as possible the stuff you throw away after you finish your lunch.

When you do this type of activity your objective is to Reduce, Reuse or Recycle everything involved in making your lunch. When you do all this you will help prevent pollution, conserve natural resources, save energy, reduce the need for disposal, and add to your goal of being a good environmental steward (humans are considered stewards (a steward is a person responsible for taking care of something and in our case as humans we are responsible for taking care of this planet) now and for the future. The following steps cover a lot of what you can do:

1. Avoid excess packaging by buying larger amounts of the things you eat and put smaller amounts you use for lunch in reusable containers.
2. Take a cloth (preferred if you can) or metal lunch box that you can re-use.
3. Put your drinks in a plastic re-usable bottle.
4. Use plastic or metal utensils that you can re-use.
5. Put your sandwich in a re-usable plastic container.
6. Replace fruit cups with the real fruit like a banana or apple and bring home the peel or core to hopefully compost (adults will need to help you set up the composting at home but there is a lot of information on line to show you how to do it.)

1. If you do have to bring an aluminum can or paper make sure is recycled but your overall all goal to be able to re-use as much as possible.

   Go to the following EPA web site for a lot more information on how to do this activity and maybe even consider it as a school project and get others involved.
   https://www.epa.gov/students/pack-waste-free-lunch

I understand you may not be the one that packs your lunch but I hope you will ask the adults in your life to help you accomplish this great activity that could really impact the environment by saving hundreds of items from our landfills by not being disposed of the way so many people do it.

Donate Toys And Clothes:

Kids Age Range – All Can Participate, But Will Need Adults Help But A Great Family Activity

This activity may not be an obvious environmental activity to you as others but it is for a lot of reasons besides the great positive thing you do when donate your old toys and used clothes to someone or organization that use them. Think of how nice it would be for others to get things like toys and clothes that you no longer have a need for and maybe not as fortunate as you are. Sometimes it is not that another kid may get use from the toy right away but an organization or church group that may need them for when they have kids to take care of so while they are there they have some toys to play with. Clothes are especially helpful for those kids that don't have a lot or just to save some friends money that are younger than you and you can't wear something anymore because you have outgrown it. It just mean you give them to people who can't afford them because everyone hopefully appreciates the ability to save some money as well as help save the environment.

    This activity has a significant impact on or Natural Resources that are materials or things that people use from the earth. We identify these natural resources as two types. The first are renewable natural resources because they can grow again or never run out. The second are called nonrenewable natural resources which are things that can run out or be used up and are typically from the ground.

    As to the toys and cloths you may have wooded toys that are from trees which are renewable and all things usually take some amount of water to make which is also renewable but like trees it is a delicate balance and you can use so much it no longer can be renewable or depending on the area you take them from can have a real bad impact on the rest of the environment and our lives because we need trees to generate air to breath and water to live. We usually think of the renewable resources as something that regrow or be replaced within a person's lifespan.

The second type as indicated is nonrenewable natural resources that are impacted by having toys and clothes. Nonrenewable resources are typically found in the ground. Nonrenewable means there is only so much available and is usually called a "fixed amount" so if we use it up we won't ever have any more like minerals used for making metals and oil (fossil fuels that include natural gas, coal) that don't regrow and they are not replaced like water or renewed like trees. They include the fossil fuels we burn for energy (natural gas, coal, and oil). Minerals, used for making metals, are also nonrenewable natural resources. Different from renewable Nonrenewable natural resources are things that take longer than a person's lifespan to be replaced. In fact, they can take millions of years to form like oil that took the dinosaurs dying millions of years ago to make the limited amount of oil we have.

We use a lot of both types of natural resources to produce those toys and clothes we have but along with using the natural resources the process in making all this stuff increases the pollution that then hurts the rest of the environment and all the living things in our biosphere that I mentioned above. This is just one way to make sure we use all natural resources wisely. Conservation is what we do by not using up, spoil, or waste things.

When you think about what you want to do with your old toys and clothes consider some of the options below. The adults in your life will need to help make sure you are making the right decision about your toys and clothes but it is important that you try to understand why this activity is so important and I know sometimes it is tuff to decide.

Toy And Clothes Donation Options: This is just a small list of examples and are a lot in whatever place you live.

- Charity thrift stores such as Goodwill and The Salvation Army accept toys and clothes then put them on the store floor for others to buy or give them away.
- Local Shelters and Children's Centers.
- Preschools and Nurseries.
- Church Charities.
- Toys for Tots.
- Ronald McDonald House.
- Loving Hugs or Stuffed Animals for Emergencies

This is a favorite of mine because they offer children a familiar object in times of emergency. Both Loving Hugs and Stuffed Animals for Emergencies give stuffed toys to kids in crisis situations, whether it's a natural disaster or while taking an ambulance ride. Especially when you have stuffed toys because you probably recognize yourself that when you are scared or sad how much comfort a stuffed animal can provide

Garage Sale Option:

You also can actually make money in this effort by having a garage sale of your own or maybe getting some of your friends to join in with their stuff then have some extra money to save or buy other things.

By doing any of the options above think about the activities below that no matter what kind of waste (toys and clothes you don't need or want are called a waste when you decide to throw them away) you have.

Reduce:

We all generate waste and if we can minimize all that we generated by doing this and a lot more activities we can reduce the amount of waste created in the first place which is a real good thing because reducing waste saves energy, reduces the waste that ends up in landfill (place in the ground we put waste that is there for a long long time and saves those natural resources we talked about above.

Reuse / Repair:

Reusing items is what happens when we give them to someone else for them to use rather than throwing them away. Also you can repair items like toys and clothes, you may need help to do this but it is something to consider even if you are still going to donate it or sell it.

Recycle:

As to toys and clothes it is also possible, if none of the other options above can be done you can also try to have them recycled which means that the stuff can be chopped up or taken apart and put back into the process for making new things. Especially toys because a lot are made of plastics that can be ground up and the different materials in them reused. Clothes can also be recycled but it is a little more difficult but is an option. But any recycling reduces pollution and saves energy while slowing down the rate at which non-renewable resources are depleted which is a big deal.
So now you have a great activity for what to do with those toys and clothes you don't need or want and not only will you make a difference buy helping protect the environment but actually make someone's life a little better.

Monitoring Your Water Usage:

Kids Age Range – All Can Participate, But Will Need Adults Help, But A Great Family Activity

This one is for those of you that like a little technical challenge and enjoy some data collection to really see how much water you use as a family then the real challenge comes in figuring out ways to minimize this valuable natural resource. When you minimize your water use you save the water for other things like growing things, drinking, and when your water supply is from water on the surface you actually increase the habitat available for all things that live in the streams and rivers and depend on that water source. Also, you can actually save money on your families water bill especially if you minimize your hot water you can actually save even more on the cost to heat the water. So here is the activity, you want to start by doing a chart to see the amount of water you and your family are actually using at home for one week. You want to make a chart that includes the estimated water use for common daily activities, such as brushing teeth and doing the dishes. When you get a week of data then you can identify which activity needs some work to get you the most reduction in use. After about a month you want to do the chart again to see if you actions have made a difference.

The chart should have the following: If you can it is good to use lined paper and maybe graph paper if you can.

1. Heading "Our Water Use For The Week Of _____ (Example: 01/01/18 to 01/07/18)" and put the Name of the family person in the right hand corner of the sheet.
2. On the top of the page under the title from left to right make columns with the following headings: Activity, Sunday, Monday, Tuesday Wednesday, Thursday, Friday, Saturday, Total Number Of Times, Estimated Amount Of Water Used In Gallons, and Total Weekly Gallons Used.

3. Under the Column heading going down list the following Activities (more or less can be used but try to use as many as possible): Washing Face And Hands, Taking A Shower, Taking A Bath, Brushing Teeth, Flushing The Toilet, Getting A Drink (From Faucet), Cooking A Meal, Washing Dishes By Hand, Running A Dishwasher, Doing A Load Of Laundry, Watering The Lawn, Washing The Car At Home, and at the bottom Total Weekly Gallons.
4. Now finish the chart by drawing lines across from right to left and up and down to make the chart as neat as possible for your entries.
5. The next step is collecting all your data. You can either collect data from just your own use or hopefully have more people in your family help by letting you know when they do any of the activities on the list. Keeping one sheet for each person. BUT only enter gallons for Running A Dishwasher, Doing A Load Of Laundry, Watering The Lawn, and Washing The Car At Home on one person's sheet because these activities apply to all the family members.
6. When collecting the data and estimating the gallons it would be tuff to collect the water in the sink or tub or washers or calculate the volume based on the outlet flow all is real technical stuff but thankfully some real technical people have already done it to give you a typical amount to use for each one. It is not exact but it will be close enough to really see the possible savings to your family and protection of the environment. Use the following gallons for each of the activities on your list and enter under the column Estimated Amount Of Water Used In Gallons on the line where the activity is identified. Each activity that follows has brackets () with the typical gallons to the right that you should use: Washing Face And Hands (1), Taking A Shower (40), Taking A Bath (40), Brushing Teeth (2 with water running), Flushing The Toilet (4), Getting A Drink

(From Faucet) (0.25), Cooking A Meal (3), Washing Dishes By Hand (10), Running A Dishwasher (15), Doing A Load Of Laundry (30), Watering The Lawn (300), Washing The Car At Home(50).
7. Now go collect your data and if you miss a few times it is not a big deal but just try to get as much as possible. Collect all you can.
8. The number of times for each activity has no right or wrong number but it is all about what you or your family can do touse less water. Of course you never want to leave water running after use or have leaks, even small ones, that continue without being fixed but that is something an adult will need to help with. A leak of one drip per second can cost $1 per month. That may not seem like much but from the American Water Works Association who put it into perspective: at "60 drips per minute, you waste 8.64 gallons per day, 259 gallons per month, and just over 3,153 gallons per year". That is a LOT of good, clean water just going to waste! So if you see a leak you should always tell an adult right away.
9. When you get done collecting all your data add up the usage for the week.
10. Now the real tuff part starts because you need to see what you can do yourself or get help from other family members to change some of your habits to make a difference. You and any other people who you collected data need to be most involved because you are going to check again on their usage in a month to see what savings were made. Below is a list of things you can do for each activity on your list.

Washing Face And Hands: (1 gallon per event)

You need to make sure you keep clean but you can try to use less water by soaping up without the water running then you rinse and dry. (Soaping up then rinsing can save about ½ gallon)

Taking A Shower: (25 gallons per event)
You can use a lot less water when you take a shower so like when you wash up if you get wet then turn the water off then use the soap and then wash off and take showers instead of baths. One thing you will need to get adult help with is to install what is called a low-flow fixture. The laws require new showerheads and faucets to have low flow rates made after 1992 that can actually cut the amount of water in half from 50 gallons to 25 gallons, a real big savings just by itself and if you shop around they can be real inexpensive. You will need adult help to see if the showerhead is labeled low flow or it can be checked by measuring the actual flow with a bucket. Go to this web site for exactly how to measure the flow rate https://www.thespruce.com/determine-faucet-shower-flow-rate-gpm-1824859. Your local hardware store should be able to help you identify specific fixtures for your needs.
(Reducing time of water flow by soaping up then rinsing can save 5 gallons per shower.)

Taking A Bath: (36 gallons per event if a standard tub)
Again you need to keep clean and if you take a bath with only a few inches of water it can use a heck of a lot of water especially if you have one of those HUGE jetted bathtubs! So really your only option is to take showers sometime with a low flow showerhead to save the water used by taking a bath. (Taking a shower instead of a bath and soaping up without water running saves 16 gallons per event.)

Brushing Teeth: (2 with water running)

Don't let the water run. Are you guilty of leaving the water on while you brush your teeth? This item is a big one as to water use. (Brushing teeth without leaving the water running can save 1 gallon per brushing.)

Flushing The Toilet: (5 gallons per event)
Because most people never look in a toilet tank you may not even know what is inside. Again with an adults help check to see what type of toilet you have and if not already the efficient new kind. One thing that can be done to reduce water is besides replacing it with a newer water efficient one an easy way to reduce that amount, even with older toilets, is to displace some of the water in the tank with a brick, which allows you to get the same flush pressure, but you use up to half a gallon less water per flush. Okay so you are worried about crumbling brick in your toilet messing it up, then you can put a plastic water bottle in toilet tank. If you take a plastic bottle or even bottles that you have stopped using since you now use home water not from plastic bottles purchased separately, put a few pebbles or rocks in it, fill it with water, and then place it in the back of your toilet, you can save up to 10 gallons of water per day and not worry about brick pieces clogging your toilet. (Replacing the toilet with a more efficient one or adding a brick or water bottles will save at least 1 gallon per flush.)

Getting A Drink: (From Faucet) (0.25 gallon per event)
This is a tuff one but the way to cut down the water use besides what you need to drink and never ever don't hesitate to get a drink when you need one is to not let the water run before or after you fill your container. (Cannot really recommend this one to reduce because doing a little flush of the line prior to getting water to drink is my recommendation but you need to track it.)

Cooking A Meal: (3 gallons per event)
This one again needs to not be reduced if it means you need water to rinse your food or to cook with but again don't let the water run while you are doing something else while cooking. (This one is another one to be careful with but if your first observations indicate water is being left to run then they stop you can deduct 1 gallon per event on the recheck.)

Washing Dishes By Hand: (10 gallons per event)
When washing dishes by hand turn the water off or fill the sink with water first so when you step away to grab dirty dishes, or find the soap the water is not running. All of those extra minutes can add up to a lot of wasted water. It only takes a second to shut it off! A suggestion you can bring up with the adults is to consider purchasing an ENERGY STAR®-qualified dishwasher that uses 31% less energy and 33% less water. You can find a lot of information on Energy Star appliances by just entering Energy Star in your web browser. (This one has real potential either by not washing dishes by hand and waiting for a full load in a dishwasher even if you don't use an Energy Star one, your savings would be 10 gallons for each one you avoid on the recount. So if you had 20 times in a month you would save 200 gallons. However if a new dishwasher is not an option then if the way you wash them changes to make sure you only have water on when filling the sink then you can deduct 5 gallons each time.)

Running A Dishwasher: (15 gallons per event)
Besides purchasing a new dishwasher that is Energy Star you should consider using your dishwasher efficiently. Wash only full loads, choose shorter wash cycles, and to save money on energy bills as well activate the booster heater if your dishwasher has one. It's commonly assumed that washing dishes by hand saves hot water. However, washing dishes by hand several times a day can be more expensive, (pay

attention to this one as it makes a good argument to buy a new dishwasher) than operating an energy-efficient dishwasher. You can consume less energy with an energy-efficient dishwasher when properly used and when only operating it with full loads. This full load issue can be problematic if just a couple of you and you don't generate many but no problem with a large family.

When purchasing a new dishwasher, check the Energy Guide label to see how much energy it uses. Dishwashers fall into one of two categories: compact capacity and standard capacity. Although compact-capacity dishwashers may appear to be more energy efficient on the Energy Guide Label, they hold fewer dishes, which may force you to use it more frequently. In this case, your energy costs could be higher than with a standard-capacity dishwasher. The previous issue is something you do need to really try to figure out.

One feature that makes a dishwasher more energy efficient is a booster heater. A booster heater increases the temperature of the water entering the dishwasher to the 140°F recommended for cleaning. If you lower, when possible, your water heater temperature to 120°F to save energy so you need this increase in water temperature to get everything clean. Some dishwashers have built-in heat boosters, while others require manual selection before the wash cycle begins. Some also only activate the booster during the heavy-duty cycle. Dishwashers with booster heaters typically cost more, but they pay for themselves with energy savings in about 1 year if you also lower the water temperature on your water heater.

Another dishwasher feature that reduces hot water use is the availability of cycle selections. Shorter cycles require less water, thereby reducing energy cost.

If you want to ensure that your new dishwasher is energy efficient, purchase one with an ENERGY STAR® label. (Water volumes don't change using the dishwasher but what does change is the number of times you use it because you are making sure it is full. You would add a 15 gallon saving for every time less than used last time you collected data. Calculating this one is a little different because you could actually have more dishwasher events than last month because you stopped doing handwashing your dishes.)

Doing A Load Of Laundry: (45 gallons per event)
Consider upgrading your clothes washer. ENERGY STAR® says that you could fill three backyard swimming pools with the water you save over the life of a new ENERGY STAR®-qualified washer. If you're replacing a washer that's over 10 years old, you can save over $135 per year. (If you change washers for an Energy Star it will have a per gallons per wash so you can use that to calculate your savings. If you keep the same washer but make sure all loads are maxed out you should see a reduction in the number of loads so you just multiply that number times 45 for your savings minus the loads counted on your last data collection.)

Watering The Lawn: (300 gallons per event)
This is a big water use item that you need to talk with the adults about and get their okay to help with. People often water more than the grass really needs to stay healthy. So be sure not to keep it running to long and only water in the cooler part of the day. Also, you can consider replacing areas with plants and grass with surfaces that don't need as much water but if you have animals that may be a real challenge but again mention it to the adults in your life for consideration. Also one item that takes a little money is to put in an automatic system that can control when water goes on and off, over time the reduction in water with an automatic system can actually pay for itself in water savings. (This one is also tuff the easiest way is just to time the watering time and divide that time into the 300 gallons which is typical so if you water for 30 minutes it would be 10 gallons per minute. When you

make a change just use 10 gallons for each minute you don't water as your savings. Because you would need to water differently depending on the season this makes it more complicated so to get the real savings you may have to collect data for a summer month then after changes collect data again in a summer month to make the comparison.)

Washing The Car At Home: (50 gallons per event)
This I know will cost more money if you are currently doing your own car washing and can actually be fun for all but the studies show that taking your car to a carwash place actually saves money because they use less water per car and often recycle or retreat the water on-site and use it on more cars. Also, a carwash has to make sure all the chemicals they use to get the car clean are handled responsibly so the environment is protected more than when you let the water drain off the driveway into your neighborhood drainage system no matter how little water you use. (This one has a water savings just by comparing the number of washes where you save 50 gallons for each time less than the first data collection but also should account for the cleaners you don't use as a savings. Add both the water cost savings and the cleaner costs to come up with savings and subtract the cost of taking it to a carwash to get a true savings. This one may really come out more expensive to take the car to the carwash but you really will be helping the environment a lot.)

Okay now that you know how to get the efficiency in your heated water management and minimize your water use pick one of the above or maybe two or all of them to start saving more money and do the planet some good.

After you have tried to do what you can and made the changes for about a month then take your measurements again to determine how much you saved by subtracting your new total from your first total. Use the statement in bold below each activity to calculate what your new use is after you make the changes.

Figure out your savings by multiplying your cost per gallon times the number of gallons saved. Typical costs for water out of the tap or hose at home is about $0.005 per gallon for community supplied water. You should ask the adults in the family to look at your water bill to see what the actual cost per gallon is. When you calculate the savings remember that you need to add the totals from EACH data sheet because they will be different and you don't want to count activities like laundry and dishwashing, lawn watering or car washing that are for everyone more than once. You now multiply the week savings in water gallons times 4 for the savings per month. So if you can cut water use by just 1500 gallons in a week, that is not unreasonable, you would save 1500 (gallons) x 4 (weeks) = 6000 gallons per month. So you would save 6000 (gallons) x $0.005 (minimum cost) = $30 per month or $360 per year which is a lot to have available for other things not even counting the positive impact on our natural resources.

Even if you are using well water the savings could be real important because the well water would last longer because sometimes they run out and your expensive water pump would not run as long extending its life because they are real expensive to replace.

5

SOMETHING YOU SHOULD NEVER DO THAT REALLY HURTS THE ENVIRONMENT

TRASH OUT THE WINDOW OR JUST ANYWHERE:

Kids Age Range – All Can Participate – And All Can Lead By Example

This one bothers me a whole lot and a lot of other people as well and hopefully you, as most people are bothered by non-caring people who chuck stuff out a window on the road or in a parking lot to get it out of their world and into all of ours including everything else that lives in our habitat and the larger biosphere.

I often wonder when I see someone do this intentionally, there is unintentional littering but a small percentage. I even worry that when people of all ages have an attitude so uncaring and obnoxious that they think it is okay to dump their trash out into the environment. Or is it they just don't care about what they are doing? It's not like garbage cans don't exist everywhere. This littering has always struck me as one of the most stupid things earth dwellers do in the United States and other developed countries and has bothered me since I was old enough to understand the damage. If this sounds judgmental, you're right because when things we do as people are just wrong other people often suffer along with other living things on this planet. I contribute a lot of bad things people do related to the environment, trying to be positive, is more of a lack of knowledge or skill set, this isn't one of them.

You should know that in many under developed countries they don't have any real functioning waste management systems and it is a problem at all levels that contributes to disease and other bad things with no easy fix.

What I am hoping is that anyone reading this who is doing it even occasionally will stop and more importantly the adults make sure the kids they influence in their life are aware that this is bad on so many levels. But, even as a kid who is old enough to read this you can also be an example to other kids, and adults in their life.

To Understand the Problem, First Some Facts About the Damage:

The impact of littering causes harm to the environment, pets, and wildlife. Birds, fish and other ocean-dwelling animals that are often unable to distinguish between trash and food. As a result, 1 million sea birds and 100,000 marine mammals die each year after becoming entangled in or eating litter.

Altogether, it adds up to around 52 billion pieces of litter cluttering up the planet. That breaks down to more than 6,700 items per mile. This is why we have to work together to end littering. Research shows 85 percent of littering behavior is the result of individual attitudes.

Littering doesn't just take a toll on the environment quality; there's also an economic cost. Between clean-up efforts, decreases in property values, which is important especially to the adults, and medical expenses associated with treating illnesses caused by litter, the cost adds up to $11.5 billion each year.

The most common forms of litter are food/organic material, cigarette butts, and small pieces of paper, receipts, and gum wrappers.

Litter in the United States is an environmental issue and littering is often a criminal offense, punishable with a fine as set out by laws in many places.

Litter can also clog storm-water drains and cause flooding. Food scraps and other organic items that are disposed of improperly can increase algal blooms in water, which reduces the amount of available oxygen for other aquatic life, such as fish.

If trash is sitting in water, the water becomes contaminated, and when the water evaporates whatever was in the trash in now in the air.

Litter can build up and attract insects and rodents, which bring unwanted germs and disease to the ecosystems.

About 75% of people have admitted to littering in the past 5 years. A fact you may have never of thought about is that according to a study by Green Eco Services about 25,000 car accidents are caused in the US by litter each year. 1 in 4 Texans admits to littering in the past year. A sad comment for the state that I call home for the last 23 years. This is from the Don't Mess with Texas program.

It is important that you learn to not litter at an early age because age continues to be strongly correlated with littering. As reported by Don't Mess with Texas (DMWT) literature and other studies (average numbers) two-thirds (68%) are 16–24 years old have littered in the past month, followed by 60% of 25–29 year-olds, 50% of 30–49 year-olds, and 33% among those 50 years of age or older. I put this in because I live in Texas but the rest of states are as guilty by age groups, not the same numbers in states I could find data on but similar comparisons to age groups.

Another DMWT statement is that "Nine in ten parents surveyed have told their children not to litter and 97% would pay more attention to littering if their children asked them to so you can really help by just reminding the adults in your life politely and with respect. But only one-third (32%) of parents say their child has talked to them about littering (45% among parents who are 16–29 years old and would therefore have younger children)." This statement makes more of the case for the adults to be more involved.

From the entire US perspective, despite being bombarded with media campaigns that warn about the dangers of littering, many Americans aren't heeding the message. Approximately 43% of people in the U.S. admit to littering at least once.

The fix is more attitude than anything else by choosing just not to litter. Besides the practical solution of getting or reminding the adults to get a trash bag for our cars, boat, camping, beach going, or anything you do or place you go that does not have a readily available trash can AND USE IT. Also, don't let people do it that are part of your group or family, say something and maybe even give them one or two of the facts above, just don't let it slide. As a kid you can still give some guidance and set the example by developing the positive habit of never doing it.

One great idea for you and the kids is to volunteer in your community to do a litter clean-up event. Participating in the event will not only make you feel good about helping the planet but for kids it drives home the real problem.

6

ACTIVITIES FOR THE 4-7 YEAR OLDS

The following activities are for the younger group that should be able to be done WITH an ADULT or Older Kid's help because they require scissors and glue use and they will need help getting all the materials but can be great fun and at the same time help the environment by reusing things you would be throwing away or maybe recycling, All are great activities for one kid to several in a group and with a hope that you can explain why ding these kind of things help the environment as you assist them in making them then let them play on as in the use of the puzzles or cars they can make. The chimes are nice to listen too or the turtle egg cartons can be used to play with outside and inside. An additional advantage is they are perfect activities for that rainy day when the kids need to be indoors.

Homemade Puzzles:

Kids Age Range – All Can Participate

For this activity instead of throwing away greeting cards, cut them up and make a homemade puzzle. Cereal boxes are great for this, as well.

Tools:

1. Scissors – Sharp scissors or Art Knife but adult needs to use
2. Art Knife for sure.
3. Ruler

4. Pen / Sharpie for line drawing.

Materials:

1. Cereal Box
2. Greeting Card

The process is the same for both:
1. Find a greeting card that has been sent to you that has a nice picture on the front or a cereal box that is empty.
2. With a pair of sharp scissors as they need to be able to do a clean even cut not frayed but you need to make sure the child is supervised or do the cutting for them. If an adult helps they can use a matt knife but don't have a small child use a matt knife, it is too dangerous, and if you as an older kid or an adult use a matt knife make sure you use a board underneath as you cut to avoid any surface damage you are working on.
3. Cut away the picture front from the greeting card or cereal box being sure to cut away the picture as straight as possible and any curved edges that connect the front picture to the rest of the card or box are gone so the pieces fit flat on any surface. Now you have a choice to turn over the picture showing the backside and either take a flat edge and line out the back in any shapes you want but I suggest you don't do circles unless you are using a matt knife and use a bottle top or something to first trace the circle before you cut.
4. Now cut the picture up closely following the lines you just drew. Of course you don't have to draw any lines and just cut various shapes but depending on the child's age small pieces can be a problem and you want to keep this puzzle stored for more play time and small pieces get lost easy as all of us adults know.
5. Now just have fun.
6. Below are the step by step pictures for the process using a

greeting card and cereal box.

Materials: Scissors, Cereal Box Cover, Greeting Card, Black Marker, Ruler (you can use any straight edge)

Puzzles Ready To Cut Up And Use

Assembled Card Puzzle

Assembled Cereal Box Front

Homemade Wind Chimes:

Another great activities for cans you can recycle but it is better to reuse as wind chimes that can really be an easy project giving lots of lasting joy. The materials and process are found below that I did but a great video from Guidecentral English on a version of the project is found at https://www.youtube.com/watch?v=IiUtqC4SvLs.

Tools:

1. Metal Punch or Large Flathead Nail
2. Hammer
3. Paint Brushes For Acrylic Paint
4. Pen or Sharpie to mark twine / string

Supplies / Materials:

1. 3 Steel / Aluminum Cans of different sizes.
2. 3 feet of twine / string
3. Acrylic Paint (suggest four colors but you can uses two with various color schemes)
4. Stickers or any small metal decorations that you can attach securely to the cans.
5. Ring of metal / brass to tie the string end to so it can be hung.

Process:

1. Select 3 cans of large to small with all cans fitting inside of the can of the next size.
2. Clean all cans thoroughly and peel off all labels.
3. Punch hole in top of can you still have that will now be the top of the can you will attach.

4. Paint all cans and decorate as you would like on the outside. Stickers, painting patterns, Tacks or other metal stick on decorations. Suggest you don't glue anything other than stickers to outside of cans. If you use metal decorations attach to can with screws or tacks but bend down the edges inside the can.
5. When paint is dry attach all decorations.
6. Run twine or string through hole in smallest can from inside out and tie large not in the end that will be inside the can stopping the twine from coming out.
7. Rune the untied end through the next size can and position can so it hangs about two inches over the smaller can below it and mark the spot with a line (Sharpie marker works best for this) take string out of second can and tie not at marked point (close is good enough here it does not need to be exact). After tying knot restring the can so it hangs somewhere close to where you marked its position.
8. Follow the same steps for cans 2 and 3.
9. With all cans now stung each can should cover the can below by about 2 inches.
10. After all cans are strung cut your twine to length you want it hang bellow from where it will be placed but suggest not more than one foot to prevent it swaying to much in the wind.
11. Tie metal / brass ring to the end of the twine and hang it up, you are done. Just make sure it is high enough to prevent little kids from pulling it down.

Materials: Twine to tie cans together and hang, Paint, Brushes, Scissors, Tacks to attach to cans on outside, Ring to tie twine end for hanging, Punch, Hammer.

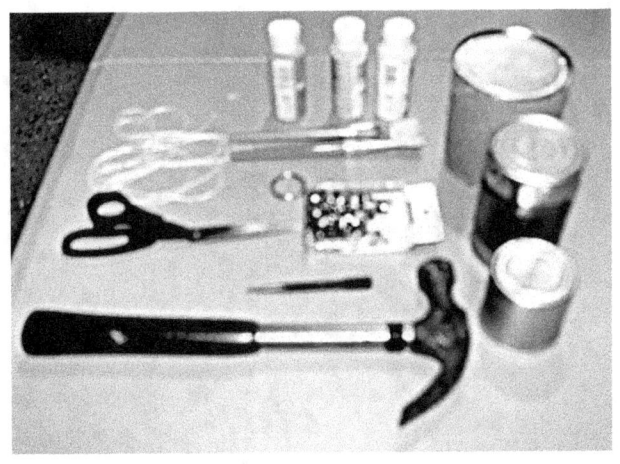

Chimes Ready For Assembly And Final Decoration

Assembled Chimes And Hung Up Swinging In The Wind

Toilet Paper Roll Car Craft:

This activity can be a lot of fun for several kids and the adults. When you are done with a little care they can be a toy(s) that will last for lots of play time. I will go through the steps below but for a great web site that also has directions and pictures go to https://www.handimania.com/diy/toilet-paper-roll-race-cars.html provided by Handimana.

The process and things you will need are are as follows:

Tools:

1. Scissors and in this case a Craft Knife or Matt Knife
2. Hole Punch
3. Glue (glue stick works but bottled glue with applicator brush seems to work better)
4. Small paint brushes – kid sizes.

Supplies / Materials:

1. Toilet paper rolls – suggest all the same size. At least 2 but if possible 3 or 4 to make 3 cars and have enough just in case you make a mistake. Also, you can use a wrapping paper roll that is cut up in 6 inch lengths.
2. Acrylic paint (water base) and small containers to put small amounts of paint in for the kids to help paint and not get the whole paint container mixed up. You should have one color for each car but if you only have two colors you can mix or put designs on them to make each one different.
3. Cardboard to cut wheels from – a cardboard box cut up in flat pieces works great – thicker cardboard the better that will make a stronger wheel but really any box cardboard will do. You need enough cardboard to cut four wheels out for each car based on the size of wheel. You are going to use a top of a jar to trace out the wheel so get the jar top you are going to use first to determine how much cardboard you will need.
4. Stickers to decorate the cars but you can glue on colored paper if you want.
5. Small prong fasteners also called brads to connect each wheel to the car. Brads are two-pronged paper fasteners, usually with a decorative top. Brads add moving elements to your paper wheel when two pieces are fastened together.
6. 3 small buttons (one half inch wide at most for steering wheels)

Process:

1. Cut out a driver's spot about 1 inch on three sides with the side not cut being the driver's seat. Fold down the flap like an "L" so you have a seat shape.
2. Paint three of the paper rolls and get creative.
3. Let all car body you just painted dry completely before

going on.
4. While waiting for the cars to dry paint the four buttons for steering wheels and let dry (if you have black buttons they don't need painting).
5. Put a clear large number on top of each car where the hood of the car would be. Let numbers dry before going on.
6. Apply stickers.
7. Using jar cover about one and one half inches wide trace out 4 wheels for each car and cut them out with Art Knife or sharp scissors.
8. Paint wheels black or any other color, the same color for each set of 4.
9. Using a hole punch make a hole in center of each wheel and another hole in the four spots on the car wheel wheels will be attached.
10. Put the Brad through the wheel and the hole in the car will the wheel will go and bend out the inside part of the Brad that is in the interior of the car but make sure it is not so tight the wheel won't turn easily.
11. Attach the steering wheel by cutting the sides where the steering wheel goes about one eight of an inch and gently folding down the edge to glue on the steering wheel in the center of the flap.
12. You are now ready to RACE!!!!

Materials: Toilet Paper Rolls (Cars), Paints, Glue, Black Buttons (Steering Wheels), Wheel Trace Guide (Bottle Cap), Brads (Spokes For Wheels), Brushes, Craft Knife, Hole Punch, Matt Knife, Cardboard For Wheels.

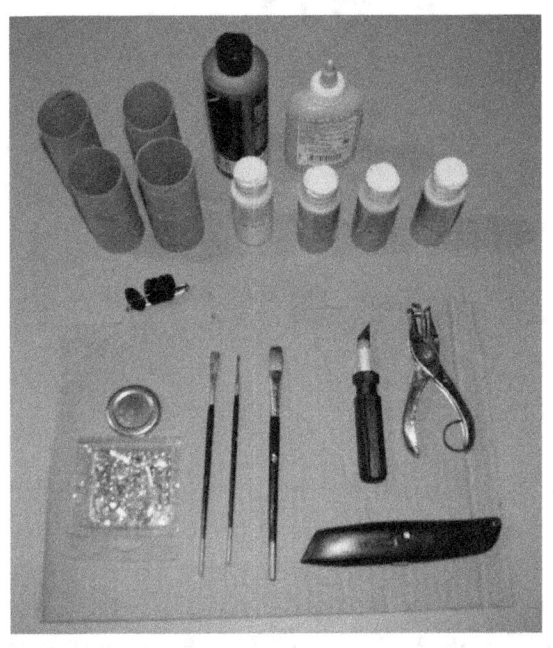

Cars Ready For Assembly And Final Decoration

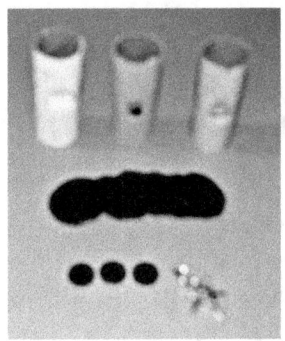

Cars Put Together With Some Mickey And Spider Man Stickers

EGG CARTON TURTLE:

This activity is an environmentally sensitive way to use recyclables to make a very attractive toy turtle. And who does not like quite turtles. I will cover the whole process below but for a good site to go to see http://www.emmaowl.com/egg-carton-turtle-recycled-kids-craft/#_a5y_p=5038998 from Emma Owl that will also take you through the process.

Tools:

1. Scissors

Materials:

1. Egg Carton bottom (shells). Green or Gray suggested.
2. Acrylic Paint – Green for the shell and gray for the legs and tail.
3. Pom Pom heads – small green or gray.
4. Eyes small flat kind that can be glued to the pom pom as part of the head.
5. Sheet of cardboard to cut out 2 of the templates below.
6. Tracing Paper
7. Glue for attaching shell and head.

Process:

1. Cut out shells.
2. Paint shells and let dry.
3. Trace out feet and tail from template from the Emma Owl web site referenced below, they also have a link to print a PDF copy if you would prefer. Other option is to trace out the pattern using tracing paper found in this book below. Pick the size that better shows the shell from the two options. (one small and one large based on egg carton size) Template From Emma Owl at http://www.emmaowl.com/egg-carton-turtle-recycled-kids-craft/

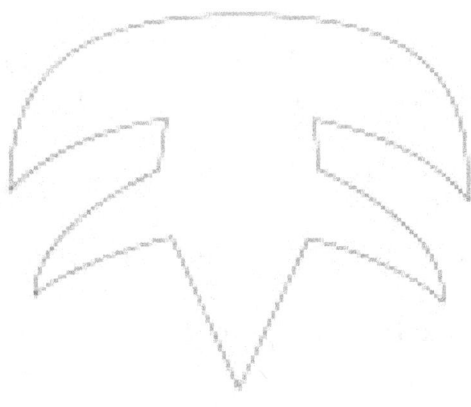

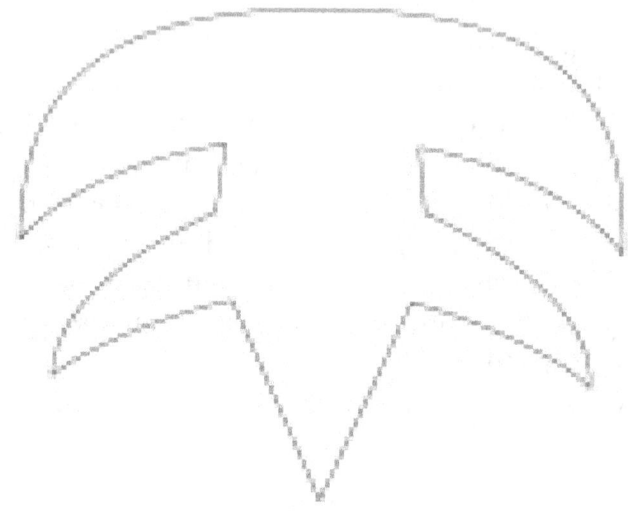

4. Cut out the feet and tail template following your traced lines on your tracing paper on top from the sheet of cardboard.
5. Paint the feet and hands cutout gray. Let dry before going on to the next step.
6. Glue the body to the template with the pointed edge being the tail.
7. Glue the eyes to the pom pom head and attach to the body.

You're done and ready for some I am a turtle time. A great time when complete to talk about Nature Conservation and how protecting the sea turtles are so important. A good site to go to is the Ducksters Education Site for more information for you and kids at this age found at the following: https://www.ducksters.com/animals/sea_turtles.php

This picture below is of a Loggerhead Sea Turtle (photo credit David Mark) – it shows you the paddle like arms and legs...

Materials: Glue, Paint Brushes, Pom Poms, Paint, Egg Carton Sections (Shells)

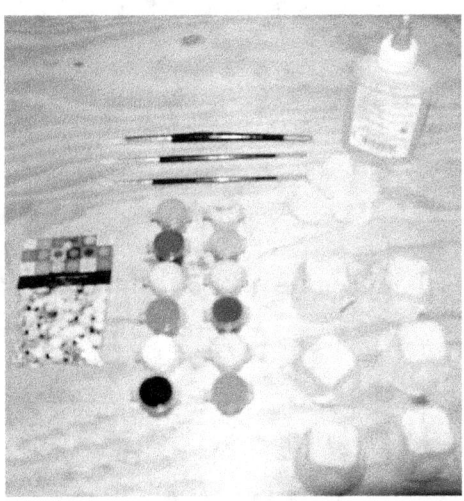

Materials: Scissors, Marker, Cardboard With Traced Turtle Bodies

Turtles Ready For Assembly And Final Touches

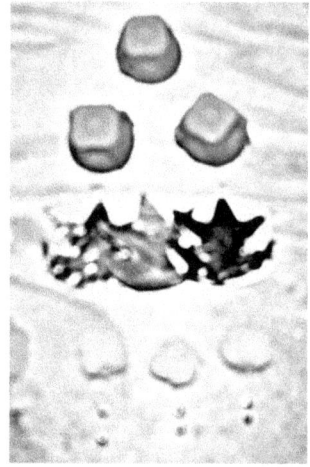

Turtles Together And Playing In The Grass

7
ACTIVITIES FOR 7-TEENS

Make Recycling Bins For The Home:

Kids Age Range – All Can Participate – And All Can Lead By Example

This activity is pretty basic but can be a good activity for you to get everyone involved in the recycle process buy making small bins that can be kept at various locations in the house that hold recycle materials. You can get creative in your decorating and labeling of the containers but there are no real rules just make sure the container is strong enough to hold the materials but they can be made of plastic, cardboard, and metal. Below are a few videos, lots more on the WEB if you want even more, to give your ideas on the project. Suggest you start with the first one listed for a review of what is recycling and maybe have other kids view it with you. The other examples listed show a variety of ways to make the bins and provide all sorts of interesting ideas but there are no real rules so use your imagination and have some fun with it.

Once you have the smaller collection containers in the house in places where they can be collected easier then hopefully getting the recyclables to the larger bins for pick-up should be a lot easier and get everyone involved.

What is recycling - Recycling facts for kids - learning for kids - Simply e-learn kids:
Available from Simply E-learn Kids, Published on Apr 13, 2017 at
https://www.youtube.com/watch?v=0Xat8b3xGSs
Aluminum Can Recycle Container:

Available from PepsiCo Recycling, Published on Sep 29, 2017 at https://www.youtube.com/watch?v=POWbXJ_svgU

How to a Recycle Bin Cart and never again carry that mess out to the street:
Available from Briggs Repair, Published on Sep 24, 2011 at https://www.youtube.com/watch?v=_YvI61KbIfE

How to make a recycling bin out of bottles:
Available from Pasha, Published on Feb 12, 2017 at https://www.youtube.com/watch?v=uimTYG8uFHg

How to make recycled trash bins:
Available from ZoomTV.sg, Published on Aug 29, 2015 at https://www.youtube.com/watch?v=_FVSZt1cHSw

Cardboard Recycle Bin DIY Projects
Available from DIY Projects, Published on Feb 15, 2017
https://www.youtube.com/watch?v=_R4htfzUJRE

Recycled And Planting Containers:

Kids Age Range – All Can Participate

This activity not only can be a lot of fun for everyone but save some money by using cans you would send for recycle. Reuse is always better than recycle as it helps the environment by not using up natural resources to make a new container even when the process uses recycled aluminum from cans you could recycle. The process is below with some suggested things you can plant in the new planters you are going to make.

Tools:

1. Hammer
2. Punch or Large Nail

Materials:

1. Aluminum Cans (Number is up to you and the sizes also your choice)
2. Potting Soil
3. Seeds or Sprouts for the things you would like to plant (see listing below of possibilities)
4. Paint (Acrylic) or Stickers or leave the labels on.

Process:

1. Collect aluminum cans that you would normally recycle of any size.
2. Clean inside of the can thoroughly and remove all labels.
3. Punch three holes in the bottom of each can with the hammer and a punch or use a large nail. Be careful with the punch and nail hole making process.
4. Paint the outside of the can or use stickers or leave as is,

anyway you would like, get creative because there are no rules here.
5. If you paint them let the can dry before going on to the next step.
6. Fill the can with potting soil to about one inch from the top and tap the container on a hard surface to compact the soil in the can a little bit but does not need to be real solid just firm.
7. Use your finger to make an indentation about one half in deep in the spots where you want to put your seeds or sprout to plant.
8. Place your seed or sprout and cover the seed with soil or push soil around the sprout if that is what you use.
9. Water the first time by pouring enough water to about ¼ inch of water over the surface.
10. Place outdoors where the planter can drain on the ground or brick / cement surface or if indoors make sure it is in a planter box that can handle a little drainage from the container. It should not drain from the bottom much but just make sure if it does it won't damage anything.
11. Placement should be in a place that gets some sun depending on the type of plant.
12. If you plant something that is not only beautiful to see grow but useful like parsley or basil and many other things you or the other members of the family can use to cook. Watering needs to be done weekly unless they get rained by putting enough water to cover the inside surface of the can should be plenty.
13. The planter should last for years so if you decide to change what is planted later feel free to do so.

Bugs, Lizards, Other Small Creatures, And Plants In The Back Yard:

Kids Age Range – All Can Participate – Great Fun For All – Including Adults

This one is all about appreciation for the Natural and how just looking in your own backyard can tell you a lot about the wonders of nature and the environment you live in. This is a kid activity that needs an adult or older kid to help with and experience shows that if the kid is 7 and above you get the most willingness to see the project through and can really see their enthusiasm for the process as you guide them through it. The activity takes place in your backyard small or large and if you don't have a backyard then the local park or any green area will do. It is one I have done with my daughters and their friends. It can be both an exciting and fun experience for all and really provide some knowledge all the creatures and plants we live with and share this planet with.

   We will call this activity "Things In My Life Not Human". The activity is a survey of your back yard, park or greenspace where you document all the living things you can find and where they live in that space near you. If by some chance you have a backyard that has more than about 50 yards square then limit your exploration to only about a 50 square yard area hopefully with some of the area with trees and or shrubs or tall grasses. But if your backyard is only about 10 yards square this project will work.

Materials You Will Need:

1. A Notebook like you use in school for taking notes or about 10 sheets of paper 8 ½ inches x 11 inches not with lines stapled together in one corner.
2. Pencils and crayons. You can have a color set of pencils instead of crayons that should have at least 10-12 colors or

a regular box of crayons and one or two pencils to write with.
3. You can have a small kid friendly camera to take pictures or just rely on trying to draw all the things you see. You can buy a kid friendly digital camera and case for about $40 - $60 on line. Disney makes a lot of them and they are USB capable for download with some neat features just for kids.
4. You can buy a couple state focused references for bugs and plants but are not needed because you can find them on the internet when you help them determine what they found out there.
5. A pair of gloves for you and them (should not be more than $15 for a few pairs of gloves) because you will be turning over rocks and moving bushes out of the way to see things up close.
6. A cheap magnifying glass that gives you at least a 6x magnification. A handheld sturdy version costs about $10 or you could by a folding set that can do 5x-20x for about $35.
7. Something to keep all the stuff in for this activity like a small backpack or even an old tackle box.

What to Do Now:

1. Draw a gird (small map) of the area to explore making sure you indicate some sort of distance scale like 1" = 1' or 1" = 2' but you need to keep this grid no bigger than one sheet of paper. Identify sections of your grid as area 1-5 or use a name for the area like near dog house or barbeque or big tree, anything that you will recognize later. Use a sheet of grid paper if you would like but not needed, nobody is grading this it just needs to be neat enough to be able to identify the places where you find stuff.
2. With your grid drawn you are ready to go but be sure you

and the kids are dressed okay to go out in your yard to get on your knees and walk thought high grass and bushes if you have any. The kid can get down on their knees only you don't have to be able to but can, it is a kick to experience close up what they see something close up but standing next to them is great also.

3. With your camera in hand, magnifier, notebook close with your grid on the first page either on the first page of your notebook or glued in grid paper glued onto the first page, and in your "field" gear you are ready to go.
4. What you are going to be doing is actually a modified Nature Field Journal that will document all your observations over time in the kid's words and with your help in listing names of things, pictures or drawings so make sure the pages are numbered and the date and time is recorded at the top of each page. Also, please record the weather, hot, cold cloudy, rainy type stuff and estimate (or check for the actual) the temperature. Identify areas in your journal. Use one page of the for each area for now and you can add more later as needed if using 10 sheets of paper 8 ½ inches x 11 inches not with lines stapled together in one corner packet. Allow for 5 pages for each area if using a notebook.
5. Now the real exploring can start by heading out in that greenspace or backyard with as much or as little green area as you have.

- Remind everyone that they need to be a little quiet in this adventure because we don't want to scare anything — of course you may have a reason to get a little excited as you find things but try to keep the loud noise to a minimum.
- Start anywhere just make sure you identify where you are at on your map so you can write down what you found.

- Start by taking pictures of your surroundings and add any knowledge you have on the bugs, plants, and other creatures they are viewing where you are standing as close-ups capturing the grasses, trees, bushes etc.. If you decided to draw everything then just do the best you can.
- Now start looking for anything living in your area, taking pictures and talking about what you find. The adult or older kid needs to jot down names of things in each area on the map in the area you see them. If something is traveling through like a bird or flying insect just note that on the map.
- Let the kid take all the pictures but encourage them to let you know what he or she is trying to capture.
- Take close-ups of all the flowers and plants for identification later.
- Any bugs or spider pictures should be taken with caution but take them.
- If you have rocks and can easily turn a few over do that but again be prepared to move back if something crawls out but always put the rock right back where you found it after you taken a picture or two. I have also turned the rocks over for the kids and had them ready to take pictures which can be real exciting sometimes, those time when something crawls out or appears under the rock as you are up close can get your blood pumping as well.
- Move bushes or high grass away gently and investigate under them and on them to see what they can find to document.
- Trees you may have also need to be investigated from the ground by seeing if holes exist for squirrels or birds as well as nests being sure to get as clear a picture as possible for checking them again later for signs of life.
- The nest issue especially as we found a Robin's nest and

actually heard the chicks and saw the momma bird making a lot of noise. Something my one daughter went back constantly to check for weeks afterward.
- Follow the same process of investigation through the whole area. Depending on the overall interest and stuff you may encounter should take you about 45-60 mins at most. The difference in the time can be just how long you stand in one area to see if things change by standing still and just observing. Observing time depends a lot on the age of the kid(s) but lots of things can happen by just sitting in the grass or standing in one spot silent.
- The last part is where your part increases as an older kid or adult unless the little kid is computer savvy and can print out the pictures or put them in a file for viewing from the camera then do the research to identify all the stuff you found. Part of the experience is doing this stuff together so hopefully that can work.
- You need to identify as much as possible and use first page in each section of the notebook or the one page labeled for each section of the stapled pages for each sections of your map you marked earlier.
- Make a list on each page with the Date and Time at the top of each listing of all things identified each day for Plants, Bugs, Birds, Other Animals and list all that you found that you will add a picture number you enter next to the entry later. What you can do at this point is print out the pictures (small versions about 4-6 per page) and add the ones that apply to each area behind each listing by stapling it to a blank page and dating each picture listing. Number each picture and then enter the number next to the listing entry on your journal pages.

When you get all done talk about all that you have found and if you have even a little success on your investigation questions will come up about how the creatures survive and will they come back if only flying by when you were out in their environment. It is important you need to know how we need to help them survive by not doing things that will harm them in how we live as humans especially make sure the little kids hear this from you as a big kid or adult. With a little luck you and them will want to follow-up and do more checking on some of the creatures or plants you identified and you know what everyone can do alone and what they need you to help with and if you don't know ask an adult. Adding to the Nature Field Journal each time by putting a new date at the bottom of the lists and adding more stuff or comments about if things are the same. Also add a new entry under your date and time for weather and temperature. Add more loose leaf pages as needed or start new sections in your notebook as needed.

All may not see a spark develop or near a complete understanding but you will have a great thing to keep to show later in life as a fond memory and I promise you and the younger kids will remember this adventure and how you care about nature. However, you never you may actually may generate the interest to keep journaling and spark an interest in nature and all its wonder.

GROUP ACTIVITIES FOR SMALL OR LARGE GROUPS OF ALL AGES WITH ADULT SUPERVISION

BEACH LITTER COLLECTION:

Kids Age Range – All Can Participate – And All Can Lead By Example – Lots Of Fun For All

Okay, so your first thought is probably Kids and Beach Litter Collection don't go together but it can be a great educational experience for all ages of kids from 5 on up with some supervision of course. You can hopefully make it more than a family event and actually get some of their friends involved for a day at the beach doing some good for the environment and great time with friends and family.

Suggest that you get a copy of a guide put out by the National Oceanic and Atmospheric Administration's (NOAA) called the Marine Debris Program Guidebook to Beach and Waterway Cleanups available from: https://aambgames.blob.core.windows.net/games-prod/media/legacy/oscar/media/beach_guide.pdf.

Another option in the State of Main area is calling the Blue Ocean Society for Marine Conservation (603) 431-0260 or visit their web site at https://www.blueoceansociety.org/. You may want to join one of their clean-ups as a group and take the group out for pizza to celebrate your efforts afterwards. Or if you can't join one of their scheduled cleanups, they will schedule a special cleanup for groups of 10 or more. Clean-ups typically last for about 2 hours, including a short introduction at the beginning about marine debris and why it's important that we gather this beach debris data. At the end of the cleanup, we'll weigh the trash and discuss what was found. They will provide all the supplies including bags, gloves, and data sheets. Volunteers are encouraged to wear work gloves to minimize our use of disposable gloves. They suggest participants dress in layers and wear sturdy, close-toed shoes for safety. Cleanups are conducted rain or shine. For more information about what to expect, please read our beach cleanup guidelines on their web site. There's no cost for the cleanup but donations are always accepted and appreciated. You can request a cleanup date or more information.

Another option in the Texas Gulf area is the Adopt A Beach pro- gram run by the Texas General Land Office to join as a group one of their clean-ups. Available from http://www.glo.texas.gov/adopt-a-beach/volunteer/cleanups/index.html

One more contact is the Ocean Conservancy for volunteering in the Gulf area and many other places available at https://oceanconservancy.org/trash-free-seas/
Lots more options out there wherever you are to join a group or get help planning your event.

Of course you can always plan and manage your own event and hopefully someone in your group has done an event like this prior and willing to volunteer to help.

Guidelines below are provided but everything is flexible and make sure you address any specific issues in your area and most important of make sure everything you do has safety in mind.

Guidelines For Your Own Event:

Event should be planned for 2-3 hours. Minimums Of What You Will Need:
1. Gloves for everyone you can get for a few dollars a pair (typically $2.50 - $5.00) at your local hardware store. Gardening gloves should work for most things. You are not going to handle hazardous materials so actual safety rated gloves should not be required but you could get one pair for an adult just in case.
2. Black / White heavy duty garbage bags. 30 gallon size that will cost about $20 for 50.
3. Trash Grabbers are an option at about at about $12.00 apiece. Obviously adds some cost if you have several kids but can double as a party gift after being cleaned for those that attend to use again and adds some fun to the process and especially for the ones that don't like to pick up the icky trash even with the gloves.

4. Two five gallon plastic buckets (about $5.00 each) for any medical waste and any lighters, canned mystery items that may be harmful (don't open containers) and none of the kids should handle these items under any conditions. They should be collected by adults and save for later disposition by authorities or if an adult knows what to do with them they can handle them later. Bring a black marker to label the collection bucket as medical waste and hazardous waste if you happen to find any. To get rid of the buckets you can call the local city or county and even the police if you need guidance. Don't let this part discourage you

because you can usually get help from the authorities if needed.
5. Make sure you bring Water, Snacks, and Sunscreen at a minimum.

What you should make sure you discuss with the kids and adults involved. You may even want to get creative and send out an invitation with all information about the event but the kind of information you need to share is provided below prior to your start:

- All need to wear appropriate attire: Remember it is always cooler at the beach. Cleanup participants should wear closed-toe shoes, such as hiking boots or sneakers, or sturdy sandals with an ankle strap. Also, dress in layers. Prepare for wind and for the temperature to be about 10- 15 degrees cooler than it is inland.
- One thing you need to do for all kids is to make sure their parents or guardians are okay with them participating in the clean-up (called a waiver for the legal types). One of the reasons to send out an invite is to make sure that they know what is going to happen and letting the kid come is giving their okay. Hopefully they can come with them. If you want to you can go to one of the sites above and get a copy of the consent forms to use it as a guide.

Remember no legal advice given here in any way shape or form. If you go to an organization sponsored event all minors are required to have a waiver signed by a parent or guardian.

On the Beach: Review All These Items As A Group

1. Do not pick up sharp items.
2. If you feel unsafe picking up an item please bring these to the attention of an adult.

3. Tell them NOT to pick up syringes, needles, any sharp objects, condoms, tampons, waste materials, or anything that looks like it comes from a hospital and call an adult. (You should decide to pick it up as a leader and manage with the 5 gallon buckets identified above or mark the area with a ring of stick or stones and notify the Beach Authorities or lifeguard.)
4. Broken glass is prevalent. These should not be picked up by cleanup volunteer kids but adults only.
5. Pile up any wood found on the beach but don't remove it, the authorities need to take care of it.
6. Do not remove lobster traps or buoys from the site.
7. If you see dead animals, look, but don't touch! If you see a marine mammal tell an adult.
8. All need to respect for the Natural Environment: Please emphasize respect the environment you're in. Try not to disturb living animals/plants. (If there are dunes and beach grasses on the beach, please emphasize they stay off of dunes and beach grass as these are part of a fragile ecosystem. You may have to do a little scouting on your own first to point out to the kids and adults the sensitive areas or get help from the local authorities if needed.)
9. Let them know where the collection point is. (Put all items into plastic bags and store them at one collection point that can be picked up by local authorities but don't use the trash cans that may be in the area. If possible put recyclable items in separate bags but that is not always possible unless you can take it home for your neighborhood recycler.)
10. Tell them to walk to the boundaries of the area you are going to start with prior to picking up anything. They should cover the whole beach from the water up to the rocks/plants. (It is usually most efficient to walk to the far end of the site without picking up trash and then pick up trash on your way back, , so that you're not carrying a

heavy bag in both directions. If you are a large group, spread out and walk in a straight line, doing a "sweep" of the beach.)
11. Please remind them to Look CAREFULLY for trash. It is really important to pick up even the smallest piece of plastic or Styrofoam. The objective should be quality of the cleaning for the area not quantity as even a very small piece of trash can impact the birds and other life forms on the beach if eaten.
12. Don't pick up Natural debris (algae, kelp, driftwood, shells, etc.) is part of the ecosystem system that we are working to restore and should be left alone. Please avoid disturbing plants and animals.
13. NEVER EVER pick up dead animals or attempt to move an injured animal — call the Beach Authorities or a lifeguard.
14. Remind them again to NOT pick up syringes, needles, any sharp objects, condoms, tampons, waste materials, or anything that looks like it comes from a hospital and call an adult.
15. DO NOT clean in any flowing storm drain outlets.
16. DO NOT pick up any weapons. Notify a Beach Authority or lifeguard and keep kids away.
17. DO NOT go in any locations that appear to be unsafe.
18. Divide into teams of 2-3 per team and remind them to stay together period — you need to have a responsible older kid or adult who can be in charge in each group. Make sure everyone is in site at all times and you as the organizer should be doing nothing other than watching everyone.

Consider yourself the Safety Officer and let them know you will be watching and to call for help if needed. One suggestion is to issue toy whistles to everyone, assuming the group can understand that they are not to be used unless really needed, that can be brought real cheap on-line or at a novelty store and lots come with a lanyard to where them around the neck.

Okay you have the information again and now go for it with the kids and have a great time saving the planet and maybe even providing a real example of what littering does to this planet and the places where we live.

Go On A Recycling Scavenger Hunt:

This activity is similar to the one above and you need to follow all the safety requirements and make sure there are informed teenagers or adults to help and make a list of the types of things to find, besides collecting everything you can. Items can include an aluminum can, plastic bag, wood, other metals, paper bag empty and paper bag with other things in it, straws just to name a few. Determine a time limit such as 30 minutes and identify someone to be the judge and time keeper to determine if the items count on the list. Now determine what the winning team or person gets such as picking where the group where eat lunch or dinner or even come up with small prizes to give winners and if you really get in to it give out ribbons. It can be really competitive but lots of fun just be careful and dispose of all the stuff you find correctly.

Plant A Garden:

This activity can be as large or small as you have energy and motivation to do. It is a tremendous way to watch how your food grows but can even offer a great opportunity to have some real tasty food to share with the family or at least have some great spices to cook with, and if you don't cook it is not a problem because those that do in your house will really appreciate what you are doing. This activity will take a little money for materials but not much and with the exception of a bag of good potting soil and the seeds, and s few nails you may be able to recycle some wood that is laying around just waiting for a good use. I will explain what you need to do below to build and plant your vegetables and or spices in a small garden box or a window type planter or if you would prefer you can use a few large flower pots if they are available.

    A site I suggest you go to for some information, there are several great ones but this is from Green Side Up with an article called "14 VEGETABLES TO GROW IN A SMALL GARDEN", MAY 11, 2013 available at http://greensideup.ie/vegetables-to-grow-in-a-small-vegetable-garden/

Tools:

1. Hammer / Nails (If you are going to build your planter, I will show you a few examples that are at reasonably costs, but really hope you can recycle some wood.)
2. Hand Rake or Trowel (It is a small tool which you use for digging small holes moving the soil around.)

Materials:

1. Containers

Container 1: Suggest you either purchase or build a planting box approximately 12" high 30" length 14" wide. Use plywood or particle board. An example from Earth Box at https://earthbox.com available 02/06/19 for about $32.95 plus shipping and is a gardening system that is real neat. From the site: "The EarthBox® Gardening System is a unique, sub-irrigated planter that allows the average home gardener to grow delicious fruit, vegetables, and herbs without having a traditional, in-ground garden. Grow anywhere, including balconies, porches, rooftops, and even indoors. Designed by commercial farmers, this container growing system is easy to set up and maintain. Just add potting mix, plants and water - follow our simple instructions, planting guides, growing guides and enjoy homegrown veggies – right at your fingertips. There is truly nothing else out there like the EarthBox® Gardening System." For another $20-$30 at the same site you can buy the "Ready-To-Grow Kit includes potting mix, fertilizer and moveable casters. Follow our instruction manual, add plants, then water and start enjoying homegrown veggies – all at your convenience. You'll be amazed at your success." This kit also provides potting mix and fertilizer so you would not have to buy those materials separately when figuring out your total costs for the project.

Note: As always I don't promote or gain from any recommendation of anything but always want you to see examples for what I suggest you do and you decide.

Container 2: Example from Creative Landscapes at http://siliconkarne.com/9-cool-small-garden-box-plans/ made from 1" x 12" boards.

All Container Rules:

Drainage is required no matter what kind of container you choose for your vegetable garden, it should have holes at the base or in the bottom, to permit drainage of excess water. Vegetable plants especially will die if left sitting in wet soil.

Color selection is also something to consider and you should be careful when using dark colored containers outdoors because they absorb heat which could damage the plant roots. You may need to paint them if too dark and hot to touch if in the sun all day and I suggest if you can't paint them to at least move the planter so it is shaded, not the plants but the pot if it is possible, and it is normally a bad idea to use metal pots no matter what color.

2. Potting Soil: Available from any gardening box store or your typical Loews, Home Depot. Potting soils come in different types and do vary in price and you can expect to spend about $20 for the soil you will need. Either one of those identified below will work but after selecting the seed packet for recommendations prior to buying your soil. All-purpose potting soil is generally cheaper but you should use some fertilizer to start off with so the Potting soil plus fertilizer is a good option that has it already in the mix. The seed-starting mix is the most expensive and as indicated helps a lot if you have selected certain seeds or decide to use cuttings as starter plants.

   a. All-purpose potting soil: Premixed soil for potting new houseplants or repotting plants that need larger containers. You can use all-purpose potting soil for potted vegetables, herbs and outside container gardens.
   b. Potting soil plus fertilizer: Premixed potting soil with a time-release fertilizer that feeds plants for several months to promote strong root development. Like all-purpose fertilizer, it works well for houseplants and garden plants in containers, and contains a blend of sphagnum peat moss and perlite. Works especially well for container plants and hanging baskets.
   c. Seed-starting mix: A seed-starting mix is specially formulated for seed germination and growth. This mix also works well for leaf, stem and root cuttings. This mix contains sphagnum peat moss and higher levels of vermiculite than other mixes, and provides the proper medium for growing seeds and cuttings quickly.

3. Seeds (Below is a listing of possible things to plant but remember if you grow root vegetables you will need at least ten inches of depth because their roots grow deeper as potatoes do unlike other things. Lots of possibilities and this is only a partial listing. Remember to read the back of the seed package for some more guidance on a specific seed. Costs for seeds typically can be about $1.25 to $2.00 a packet which is all you will need of each. Seeds are available at a feed store or in most Loew's and Home Depot, and of course on line such as Gurney's Seed & Nursery Co. at https://www.gurneys.com Normally you would only plant 3-4 items per garden box of the size recommended above.

Tomatoes: Cherry tomatoes only that don't require a lattice to grow on.
Potatoes: You can allow store-bought potatoes to grow "eyes," cut the potatoes in half, and plant them in a container with a depth of at least 10 inches. Make sure they will be in a place where they'll can get plenty of direct sunlight.
Cucumbers
Carrots
Peppers
Green Onions
Radishes
Turnips
Beets
Brussel Sprouts
Green Beans
Strawberries
Squash: Takes a lot of space in the garden but are great and usually you would only grow it by itself in a single box as they require a lot of nutrients and could impact the growth of other things in your garden box.

Chives
Parsley
Salontra

Bazil: When growing basil outdoors, it is very important to remember that basil is very sensitive to cold and even a light frost will kill it. Do not plant seeds or basil plants until all danger of frost has passed.

Shallots: One set (immature bulb) planted in the soil will develop into five or six shallots and also store well over winter.

Garlic Bulb: When you plant one clove, it will develop into a whole bulb.

Protection From The Possible Pests:

Now that you have your garden all planted and ready to start reaping the rewards for all your hard work you unfortunately need to be concerned about any threats from possible pests. First a good basic protection is to raise your garden boxes off the ground by using egg crates or bricks stacked securely together to a minimum of about three inches in height. If you have a chance of rabbits, raccoons, and gophers you will have to surround the garden box with chicken wire or a short picket fence of about ten inches high and install it so it can be easily removed and set aside when you want to work your little garden. Last is the small pests like ants or onion maggots that can be stopped with some dried Chile Peppers crumbled up to dust but be careful not to get this dust on your hands or eyes then spread it around the plants but not on them.

Your Growing Season:

As you get into the garden process and start to really take an interest you, and read the back of the seed packages you will learn about the best time of the year to plant certain things in the summer and some things in the spring. One thing that applies to everything you plant is to pick what you can when it is ready to eat to keep the nutrients available for the plants still growing.

Things like chives can grow to about 2 feet tall and 1 foot wide and should be trimmed regularly to encourage a new growth. Fresh chives can be dried or frozen for later use. Cut the chive plant down to 1 or 2 inches above ground level, using a pair of sharp scissors or a knife being careful and they will grow back, parsley need trimming that increases the yield of the plant. Also If parsley is not occasionally thinned, it loses vigor and if you don't cut it back occasionally it could take over and choking out other plants, and basil you will need to do trimming of the leaves from the sides and use them in the kitchen and still do a harvesting of the whole plant as needed typically about 7-9 inches high.

Wind:

Wind can be a real hazard, but we are not talking breeze because they can actually benefit the growth of your plants but strong wind can obviously blow over your garden box if elevated and actually do damage to a plant by blowing it over to far. So consider some sort of wind block or at least put some thought into where you place the box in consideration of any possible strong winds.

Water:

The obvious thing is to make sure you water every few days to provide your garden box with one inch of water per week. The one inch rule is based on making sure there is enough water to seep down into the deep root zones for plants that need it. Here in Texas typically once a week watering is enough but your area may be a little different.

Have Fun!!!!

Now that you have some basics and some starter information on getting that garden box started make sure when you select something it will be what you want to eat or at least what you always have wanted to try and don't forget that selecting things like cherry tomatoes, potatoes, or shallots can actually save some money.

9

A REAL IMPORTANT ACTIVITY FOR KIDS AND EVERYONE "EARTH DAY"

PARTICIPATION IN EARTH DAY:

This activity is so important for individual participation as well as for teachers of all age groups to use this world wide expression of concern for this planet and the environment we all share. I have included a significant amount of information directly from https://www.earthday.org/about/the-history-of-earth-day/ because they provide it in the best form I could find but encourage you to go to the site yourself and learn more as well as see how you can get involved. For the teachers reading this it is a great opportunity to focus your class on the many activities in this book as well as others that the whole school can participate in small and large.

The History Of Earth Day:

Each year, Earth Day—April 22—marks the anniversary of the birth of the modern environmental movement in 1970. Setting the stage for the first earth day:

The height of counterculture in the United States, 1970 brought the death of Jimi Hendrix, the last Beatles album, and Simon & Garfunkel's "Bridge Over Troubled Water." War raged in Vietnam and students nationwide overwhelmingly opposed it.

At the time, Americans were slurping leaded gas through massive V8 sedans. Industry belched out smoke and sludge with little fear of legal consequences or bad press. Air pollution was commonly accepted as the smell of prosperity. "Environment" was a word that appeared more often in spelling bees than on the evening news.

Although mainstream America largely remained oblivious to environmental concerns, the stage had been set for change by the publication of Rachel Carson's New York Times bestseller Silent Spring in 1962. The book represented a watershed moment, selling more than 500,000 copies in 24 countries, and beginning to raise public awareness and concern for living organisms, the environment and links between pollution and public health.

Earth Day 1970 gave voice to that emerging consciousness, channeling the energy of the anti-war protest movement and putting environmental concerns on the front page.

The Idea for the first Earth Day:

The idea for a national day to focus on the environment came to Earth Day founder Gaylord Nelson, then a U.S. Senator from Wisconsin, after witnessing the ravages of the 1969 massive oil spill in Santa Barbara, California. Inspired by the student anti-war movement, he realized that if he could infuse that energy with an emerging public consciousness about air and water pollution, it would force environmental protection onto the national political agenda. Senator Nelson announced the idea for a "national teach-in on the environment" to the national media; persuaded Pete McCloskey, a conservation-minded Republican Congressman, to serve as his co-chair; and recruited Denis Hayes from Harvard as national coordinator. Hayes built a national staff of 85 to promote events across the land. April 22, falling between Spring Break and Final Exams, was selected as the date.

On April 22, 1970, 20 million Americans took to the streets, parks, and auditoriums to demonstrate for a healthy, sustainable environment in massive coast-to-coast rallies. Thousands of colleges and universities organized protests against the deterioration of the environment. Groups that had been fighting against oil spills, polluting factories and power plants, raw sewage, toxic dumps, pesticides, freeways, the loss of wilderness, and the extinction of wildlife suddenly realized they shared common values.

Earth Day 1970 achieved a rare political alignment, enlisting support from Republicans and Democrats, rich and poor, city slickers and farmers, tycoons and labor leaders. By the end of that year, the first Earth Day had led to the creation of the United States Environmental Protection Agency and the passage of the Clean Air, Clean Water, and Endangered Species Acts. "It was a gamble," Gaylord recalled, "but it worked."

As 1990 approached, a group of environmental leaders asked Denis Hayes to organize another big campaign. This time, Earth Day went global, mobilizing 200 million people in 141 countries and lifting environmental issues onto the world stage. Earth Day 1990 gave a huge boost to recycling efforts worldwide and helped pave the way for the 1992 United Nations Earth Summit in Rio de Janeiro. It also prompted President Bill Clinton to award Senator Nelson the Presidential Medal of Freedom (1995) — the highest honor given to civilians in the United States — for his role as Earth Day founder.

One Billion People:

Earth Day is now a global event each year, and we believe that more than 1 billion people in 192 countries now take part in what is the largest civic-focused day of action in the world.

It is a day of political action and civic participation. People march, sign petitions, meet with their elected officials, plant trees, and clean up their towns and roads. Corporations and governments use it to make pledges and announce sustainability measures. Faith leaders, including Pope Francis, connect Earth Day with protecting God's greatest creations, humans, biodiversity and the planet that we all live on.

Earth Day Network, the organization that leads Earth Day worldwide, has chosen as the theme for 2018 to End Plastic Pollution, including creating support for a global effort to eliminate primarily single-use plastics along with global regulation for the disposal of plastics. EDN is educating millions of people about the health and other risks associated with the use and disposal of plastics, including pollution of our oceans, water, and wildlife, and about the growing body of evidence that plastic waste is creating serious global problems.

From poisoning and injuring marine life to the ubiquitous presence of plastics in our food to disrupting human hormones and causing major life-threatening diseases and early puberty, the exponential growth of plastics is threatening our planet's survival.

Earth Day 2018 and Beyond: End Plastic Pollution:

EDN built a multi-year campaign to End Plastic Pollution. Our goals include ending single-use plastics, promoting alternatives to fossil fuel-based materials, promoting 100 percent recycling of plastics, corporate and government accountability and changing human behavior concerning plastics. EDN's End Plastic Pollution campaign includes three major components:

Leading a grassroots movement to support the adoption of a global framework to regulate plastic pollution;

1. Educating, mobilizing and activating citizens across the

globe to demand that governments and corporations control and clean up plastic pollution;
2. Educating people worldwide to take personal responsibility for plastic pollution by choosing to reduce, refuse, reuse, recycle and remove plastics and
3. Promoting local government regulatory and other efforts to tackle plastic pollution.

An additional resource for things to do to get ready or participate in earth day can be found at the following: 21 Earth Day Crafts and Classroom Activities Using Recycled Materials https://www.weareteachers.com/earth-day-crafts-classroom-activities/

So this activity can be one of the most rewarding events you can do in helping bring awareness to the plight of this planet and participate with other people who also demonstrate their concern and as educators you can make take such a great step to demonstrate your support and involvement in this planet's future because if we don't create the awareness needed in allage groups it will never change.

10

CONCLUSION

"We must become the change we want to see."

— MAHATMA GANDHI

NOW THAT I HAVE PROVIDED YOU with several ways to improve your life by doing things that will benefit you as the young reader or parent, guardian, or as a teacher of the children of our world that will impact our environment now and in the future and hopefully along the way learn something. Hopefully you will feel good about what you are doing now and helping to make all our futures a better place here on this earth and even have some fun and make money in some of the activities. Most of all participation or doing these activities it provides you a chance to be part of a real change this planet needs to get healthier and turn this abuse we have been piling on it for so many years. It is by far not just industry that is doing all the harm, even though they do contribute a tremendous amount that needs more government control and oversight, it has a lot to do with our lifestyles and demands we make on the material and product suppliers to support our perceived needed lifestyle that we need to take a serious look at as Earth Dwellers. No matter where we live we all can do things to help make the change by our actions and getting involved, even in just small ways.

    We need to make our opinions known to each other, to our governments, and frankly demand change and elect people that share our passion and knowledge of what is needed. What is also extremely critical, and perhaps, the most critical part of impacting a true change is using our collective purchasing power in the products we use every day both as adults and even as young people who are knowledgeable about the issues because no matter what your age you can influence the future by your actions no matter how minimal you may think they are, they are not.

One of our most impacts is as concerned environmental people do is to use market forces as we support those companies that manage in an environmentally responsible manner that includes our support in all things that involve sustainability.

This book is not intended to cover all the things you could do to change the world just to provide a few activities to start the process on an individual basis as a young person to realistically accomplish.

I believe we have to start communicating with each other with more of an effort to be understanding of what is involved in making changes in our life styles with all the complexities and real world existence in today's world. Everyone no matter what age can be part of the solution rather than part of the problem, you just need to do something as we all share one common group as Earth Dwellers.

REFERENCES

1. Biology Online, 2018, available from https://www.biology-online.org/dictionary/Biosphere
2. The World Counts, 2018, available from http://www.theworldcounts.com/stories/environmental-degradation-facts
3. Conserve Energy Future, 2018, available from https://www.conserve-energy-future.com/top-25-environmental-concerns.php
4. Treehugger, 2018, available from https://www.treehugger.com/clean-technology/20-gut-wrenching-statistics-about-the-destruction-of-the-planet-and-those-living-upon-it.html
5. United States Environmental Protection Agency, 2018, available from https://www.epa.gov/sustainability/learn-about-sustainability#what
6. Worldwatch Institute, 2018, available from http://www.worldwatch.org/
7. Advanced Placement Environmental Science reference materials.
8. Care2, 2018, available from https://www.care2.com/greenliving/5-reasons-why-people- dont-recycle-and-5-reasons-they-should.html
9. The Guardian (2002) available from https://www.theguardian.com/environment/2002/aug/22/worl dsummit2002.earth21
10. Natures Path, Things You Can Do To Save The Environment, available from https://www.naturespath.com/en-us/blog/nine-things-you-can-do-to-save-the-environment/

11. How Stuff Works, Katie Lambert, 10 Things You Can Do to Help Save the Earth, available from https://science.howstuffworks.com/environmental/green-science/save-earth-top-ten.htm
12. Bag It The Movie, Information available from https://tubitv.com/movies/317275/bag_it
13. Department of Natural Resources, 100 Ways You Can Save the Earth, available from http://infohouse.p2ric.org/ref/15/14300.pdf
14. Science Based Life, 100 Ways To Personally Help the Environment, available from https://sciencebasedlife.wordpress.com/2010/12/30/10-ways-to-personally-help-the-environment
15. One Cent At A Time, 101 Ways You Can Save Energy and Save Environment, available from http://onecentatatime.com/101-ways-to-save-energy-environment/
16. Mother Earth News, The Plastic Bag Problems, available from https://www.motherearthnews.com/nature-and-environment/environmental-policy/plastic-bag-problem-ze0z1302zwar
17. Sea Turtle Conservancy, Information About Sea Turtles: Threats from Marine Debris, available from https://conserveturtles.org/information-sea-turtles-threats-marine-debris/
18. Plastic Pollution Coalition, Why Is Plastic Harmful, available from https://plasticpollutioncoalition.zendesk.com/hc/en-us/articles/222813127-Why-is-plastic-harmful-
19. Arizona State University, The Biodesign Institute, Perils of Plastics: Risk To Human Health and the Environment, avail- able from https://biodesign.asu.edu/news/perils-plastics-risks- human-health-and-environment

20. The New York Times, 2018, Joseph Curtin (Opinion Contributor), Let's Bag Plastic Bags, available from https://www.nytimes.com/2018/03/03/opinion/sunday/plastic-bags-pollution-oceans.html
21. New York State, 2018, An Analysis of the Impact of Single- Use Plastic Bags, available from https://www.dec.ny.gov/docs/materials_minerals_pdf/dplastic bagreport2017.pdf
22. Center for Biological Diversity, 2018, available from https://www.biologicaldiversity.org/programs/population_and_sustainability/sustainability/plastic_bag_facts.html
23. Virginia Department of Environmental Quality, Recycling Aluminum Cans Makes Cents, available from https://www.deq.virginia.gov/Portals/0/DEQ/ConnectWithDEQ/EnvironmentalInformation/VirginiaNaturally/LAT2009/ LAT2009_RecyclingCans.pdf
24. US Environmental Protection Agency, Municipal Solid Waste Generation, Recycling and Disposal in the United States: Facts and Figures for 2012, available from https://www.epa.gov/sites/production/files/2015-09/documents/2012_msw_fs.pdf
25. City of Miami Florida, Interesting Facts About Recycling, available from http://www.miamiokla.net/DocumentCenter/View/379/REC YCLING-FACTS?bidId=
26. Waste-free Mail, 2018, Saving the Planet One Mailing At A Time, FAQ, available from http://www.wastefreemail.com/faq.html
27. University of Southern Indiana, Paper Recycling Facts, available from https://www.usi.edu/recycle/paper-recycling-facts/

28. Plastics Recycling Update 2018, Jerry Powell, 2018 Recycling Market Update, available from https://resource- recycling.com/plastics/author/jerry-powell/
29. US Department of Energy, Thermostats, available from https://www.energy.gov/energysaver/thermostats
30. Direct Energy, 2018, Josh Crank, How Much Can You Save By Adjusting Your Thermostat, available from http://www.directenergy.com/blog/how-much-can-you-save- by-adjusting-your-thermostat/
31. One Green Planet, What's The Problem With Plastic Bottles, available from https://www.onegreenplanet.org/animalsandnature/whats- the-problem-with-plastic-bottles/
32. US Department of Energy, 2009, 15 Ways To Save On Your Water Bill, available from https://www.energy.gov/energysaver/articles/15-ways-save- your-water-heating-bill
33. Rotoplas, 2017, 10 Uses For Collected Rainwater, available from http://rotoplasusa.com/uses-collected-rainwater/
34. Texas Water Development Board, The Texas Manual on Rainwater Harvesting, Third Edition, available from http://www.twdb.texas.gov/publications/brochures/conservation/doc/RainwaterHarvestingManual_3rdedition.pdf
35. Green Mom, Why Styrofoam Is So Bad For The Environment, available from https://green-mom.com/styrofoam-bad- environment/
36. Call 811, Do I Really Need To Call 811, available from http://call811.com/before-you-dig/do-i-really-need-call

37. Rutgers University, Adrienne Miller, Sheila Mohazzebi, Samantha Pasewark with Julie M. Fagan, Ph.D., Styrofoam: More Harmful than Helpful, available from https://rucore.libraries.rutgers.edu/rutgers-lib/38329/PDF/1/play/
38. Wikihow, How to Field Strip a Cigarette, available from https://www.wikihow.com/Field-Strip-a-Cigarette
39. US National Library of Medicine, National Institutes of Health, 2009, Cigarettes Butts and the Case for an Environmental Policy on Hazardous Cigarette Waste, available from https://www.ncbi.nlm.nih.gov/pmc/articles/PMC2697937
40. World Health Organization, Tobacco and Its Environmental Impact: an overview, available from http://www.who.int/tobacco/publications/environmental-impact-overview/en/
41. RSPCA, 2017, What is the most humane way to kill pest rats and mice, available from http://kb.rspca.org.au/what-is-the-most-humane-way-to-kill-pest-rats-and-mice_139.html
42. Dengarden, 2017, Michael Kismet, 5 Simple Ways to Get Rid of Mice Without Killing Them, available from https://dengarden.com/pest-control/5-Simple-Ways-to-get-rid-of-Mice-without-Killing-Them
43. Department of the City and County of San Francisco, 2012, Chris A. Geiger, Caroline Cox, Pest Prevention by Design— Authoritative Guidelines for Designing Pests Out of Structures, available from https://sfenvironment.org/
44. How to Get Rid of Mice, Top 3 Humane Mouse Traps Re-viewed—Pros and Cons of the Best, available from http://how-to-get-rid-of-mice.com/humane-mouse-traps/

45. How to Get Rid of Mice, Natural Home Remedies To Get Rid Of Mice, available from https://how-to-get-rid-of-mice.com/natural-home-remedies/
46. How to Get Rid of Mice, 10 Facts about Mice to Help You Get Rid of Them, available from https://how-to-get-rid-of- mice.com/facts-about-mice/
47. How to Get Rid of Mice, How Much Does a Mouse Exterminator Cost, available from https://how-to-get-rid-of- mice.com/mouse-exterminator-cost/
48. US Environmental Protection Agency, 2015, Advancing Sustainable Materials Management: 2015 Fact Sheet, available from https://www.epa.gov/facts-and-figures-about-materials- waste-and-recycling/advancing-sustainable-materials- management-0
49. Carroll County, Maryland, 2018, A guide to Waste Management & Recycling In Carroll County Maryland, available from http://ccgovernment.carr.org/ccg/solidwaste/default.asp
50. Center for Disease Control (CDC), Field Identification OF Domestic Rodents, available from https://www.cdc.gov/nceh/publications/books/housing/figure_ cha04.htm#4
51. Center for Disease Control (CDC), Ecology and Transmission, available from https://www.cdc.gov/plague/transmission/index.htm
52. American Veterinary Medical Association (AVMA), 2013, AVMA Guidelines for the Euthanasia of Animals: 2013 Edition, available from https://www.avma.org/KB/Policies/Documents/euthanasia.pd f
53. Lawn Care Academy, Complete Tree Planting Instructions, The No Regrets Tree Guide, available from https://www.lawn-care-academy.com/tree-planting-instructions.html

54. Ocean Conservancy, Trash Free Seas, available from https://oceanconservancy.org/trash-free-seas/
55. United States Department of Agriculture (USDA), Forest Service, Inventories Of Woody Residues And Solid Woos Waste In The United States, David B. McKeever, available from www.fs.fed.us
56. US Environmental Protection Agency, Managing, Reusing, and Recycling Used Oil, available from https://www.epa.gov/recycle/managing-reusing-and-recycling-used-oil
57. Child Likes, Remote Power Transfer — the end of batteries, available from http://www.childlikes.com/battery.htm
58. Raw Materials Company Inc., How to Prepare Batteries for Recycling available from https://www.rawmaterials.com/page/education/prepare-batteries/
59. Gemeinhardt, Ronald, L, 2016, Texas Used Oil Management: A Practical Guide for Environmental Compliance, Graduate Thesis, Green Mountain College
60. Energy Sage, 2018 Most Energy Efficient Appliance, available from https://www.energysage.com/energy-efficiency/costs- benefits/energy-star-rebates/
61. Department of Energy, When to Turn Off Your Lights, avail- able from https://www.energy.gov/energysaver/when-turn- your-lights
62. Texas Government Land Office, Adopt A Beach, available from http://www.glo.texas.gov/adopt-a-beach/index.html
63. Inc. Com, Bill Murphy Jr., 7 Steps to Persuade Your Boss to Let You Work From Home, available from https://www.inc.com/bill-murphy-jr/7-steps-to-convince- your-boss-to-let-you-work-from-home.html

64. Earth911, Recycler Centers Look-up, available from https://search.earth911.com/
65. Minnesota Pollution Control Agency, Changing your oil: An earth-friendly guide for do-it-yourselfers, available from https://www.pca.state.mn.us/living-green/changing-your-oil- earth-friendly-guide-do-it-yourselfers
66. Illinois Environmental Protection Agency, 2005, How Do I Manage My Used Oil and Used Oil Filters? Available from http://www.epa.illinois.gov/
67. American Public Transportation Association, Calculate Your Savings by Riding Public Transportation, available from https://www.apta.com/resources/aboutpt/pages/transit calculat or.aspx
68. Public Transit in Your Community, Fuel Savings Calculator, available from https://www.publictransportation.org/tools/fuelsavings /Pages/ default.aspx
69. Blue Ocean Society for Marine Conservation, available from https://www.blueoceansociety.org
70. USDA, US Forest Service, Forest Products Journal Vol. 59, No. 9, available from https://www.fpl.fs.fed.us/documnts/pdf2009/fpl_2009_f alk00 1.pdf
71. American Forest Foundation, Wood: A Good Choice for Energy Efficiency and the Environment, available from https://www.forestfoundation.org/wood — a-good-choice-for-energy-efficiency-and-the-environment
72. US Environmental Protection Agency, Sustainable Materials Management (SMM) Electronics Reuse and Recycling Forum, available from https://www.epa.gov/smm- electronics/summary-report-about-sustainable-materials- management-smm-electronics-reuse-and

73. Building Green, What These Forestry Labels Really Mean, available from https://www.buildinggreen.com/infographic/what-these-forestry-labels-really-mean
74. United Nations University, 2017, The Global E-waste Moni- tor 2017, available from https://collections.unu.edu/eserv/UNU:6341/Global-E-waste_Monitor_2017electronic_single_pages_.pdf
75. Georgia Environmental Compliance Assistance Program, Electronic Waste, available from http://www.gecap.org
76. Michigan State University, 2010, Why Buy Local, An Assessment of the Economic Advantages of Shopping at Locally Owned Businesses, Principal Author Nandi Robinson, available from http://www.ced.msu.edu/upload/reports/why%20buy%20local.pdf
77. Century Link Brand Voice, 2017, 5 Benefits Of Shopping Locally On Small Business Saturday, available from https://www.forbes.com/sites/centurylink/2017/11/20/5-benefits-of-shopping-locally-on-small-business-saturday
78. US Environmental Protection Agency, Safer Choice Program, available from www.epa.gov/saferchoice
79. Bob Villa, Joe Provey, What to Do With Old Paint, available from https://www.bobvila.com/articles/what-to-do-with-old-paint/
80. Grand Traverse County, Household Paint Disposal Guide, available from www.recyclesmart.info
81. Healthline, Natural Mosquito Repellents, available from https://www.healthline.com/health/kinds-of-natural-mosquito-repellant
82. PestWiki, 2018, 6 Effective Home Remedies to Kill Roaches (Naturally), available from https://www.pestwiki.com/best-roach-home-remedies/

83. American Mosquito Control Association, Frequently Asked Questions, available from https://www.mosquito.org/page/faq
84. US Environmental Protection Agency, 2017, Citizen's Guide to Pest Control and Pesticide Safety, available from https://www.epa.gov/safepestcontrol/citizens-guide-pest- control-and-pesticide-safety
85. Center for Disease Control (CDC), Healthy Housing Reference Manual Chapter 4: Disease Vectors and Pests, available from https://www.cdc.gov/nceh/publications/books/housing/cha04. htm
86. NATIONAL OCEANIC AND ATMOSPHERIC ADMINISTRATION, Beach And Waterway Cleanups, available from WWW.MARINEDEBRIS.NOAA.GOV
87. Well Organization, 10 Advantages of Buying Local, available from https://well.org/conscious-consumers/10-advantages-of- buying-local

ABOUT THE AUTHOR

MR. RONALD L. GEMEINHARDT, MSES, (Masters of Science in Environmental Studies) has 45 years of experience in the petroleum industry beginning with USAF POL Specialists for four years. With a significant Environmental Health and Safety (EH&S) compliance support background in Refining, Distribution, Pipeline Operations, and Retail. With a focused in-depth background in waste management including recycling and used oil management. Environ- mental support areas included Incident Command System (ICS) application, within the Shell Oil Company Response Group and three years on the Deepwater Horizon Incident. As a Shell Oil Team Leader responsibilities included support to Distribution, Pipeline and Retail residual management with a significant focus on Resource Conservation Recovery Act (RCRA) compliance. Primary activities include, staff management, waste management, incident management and investigation support, remediation, litigation support, plan writing and review, training development and implementation, and on-site training. Facility operational support and on-site management background in Safety, Health, Environmental Compliance related to Air, Water, Waste, Comprehensive Environmental Response, Compensation, and Liability Act (CERCLA), Toxic Substances Control Act (TSCA), site remediation, and Department of Transportation (DOT), Fuels, Used Oil collection and processing, and Finished Lubricants production compliance management. Refining and Lubricant Compounding QA/QC laboratory management. Specialized Trainer Highlights include Terminal / Pipeline / Retail Environmental Compliance Training.

www.ingramcontent.com/pod-product-compliance
Lightning Source LLC
Chambersburg PA
CBHW071413210526
45465CB00001B/366